Earth System Science: A Very Short Introduction

VERY SHORT INTRODUCTIONS are for anyone wanting a stimulating and accessible way into a new subject. They are written by experts, and have been translated into more than 40 different languages.

The series began in 1995, and now covers a wide variety of topics in every discipline. The VSI library now contains over 450 volumes—a Very Short Introduction to everything from Psychology and Philosophy of Science to American History and Relativity—and continues to grow in every subject area.

Very Short Introductions available now:

Available soon:

For more information visit our website

www.oup.com/vsi/

Tim Lenton

EARTH SYSTEM SCIENCE

A Very Short Introduction

OXFORD
UNIVERSITY PRESS

OXFORD
UNIVERSITY PRESS

Great Clarendon Street, Oxford, OX2 6DP,
United Kingdom

Oxford University Press is a department of the University of Oxford.
It furthers the University's objective of excellence in research, scholarship,
and education by publishing worldwide. Oxford is a registered trade mark of
Oxford University Press in the UK and in certain other countries

First edition published in 2016

Impression: 8

Published in the United States of America by Oxford University Press
198 Madison Avenue, New York, NY 10016, United States of America

British Library Cataloguing in Publication Data
Data available

Library of Congress Control Number: 2015952193

ISBN 978-0-19-871887-1

Printed in Great Britain by
Ashford Colour Press Ltd., Gosport, Hampshire.

Contents

List of illustrations

Chapter 1
Home

When humanity first looked back at the Earth from space, the obvious unity of the planet that supports us—and all the life that we know of—entered the popular consciousness. Earth system science is the research field born out of this revelation—it seeks to understand how our planet functions as a whole system. The scope of Earth system science is broad. It spans 4.5 billion years of Earth history, how the system functions now, projections of its future state, and ultimate fate. It considers how a world in which humans could evolve was created, how as a species we are now reshaping that world, and what a sustainable future for humanity within the Earth system might look like. Earth system science is thus a deeply interdisciplinary field, which synthesizes elements of geology, biology, chemistry, physics, and mathematics. It is a young, integrative science that is part of a wider 21st-century intellectual trend towards trying to understand complex systems, and predict their behaviour. This chapter explains how Earth system science emerged and introduces some of its fundamental concepts.

Signs of life

To see something afresh it often helps to look at it from a new angle—and it was one person's thinking about how to detect life on Mars that gave us a new scientific perspective on the Earth. The year was 1965 and James Lovelock was employed by NASA as part of what became the Viking missions to Mars. Tasked with designing a means of detecting life on the red planet, Lovelock realized that going to Mars was not really necessary. In order to stay alive, organisms must consume materials, transform them chemically, and excrete the waste products to their surroundings. The gaseous atmosphere of a planet is a natural source of materials and dumping ground for waste products. Hence, he reasoned, if life is abundant on Mars—or any other planet—it will show up in the composition of its atmosphere.

The composition of the atmosphere of other planets can be deduced from the Earth by looking at the spectrum of radiation transmitted through them—because different gases absorb radiation at different wavelengths. Soon after Lovelock had proposed life detection through atmospheric analysis, the first observations from land-based telescopes showed that Mars had an atmosphere dominated by carbon dioxide, just as would be expected in the absence of life. So too does Venus. But the Earth has a remarkable atmosphere, containing a chemical cocktail of highly reactive gases, sustained by life (Figure 1).

Oxygen is the prime anomaly—at just over a fifth of Earth's atmosphere it is essential for our existence as mobile, thinking animals, but without photosynthesis to create it, oxygen would be a very rare trace gas. Mixed in with oxygen are gases like methane that react eagerly with it—so much so that they are on the verge of combusting together. The only explanation for the remarkably high concentration of methane in today's atmosphere is that it is continually being produced by life. Carbon dioxide, on the other hand, is surprisingly scarce in today's atmosphere. The explanation for that too, as we will see, involves life.

The faint young Sun puzzle

Whilst Lovelock was thinking about how to detect life on a planet, Carl Sagan was down the corridor at the Jet Propulsion

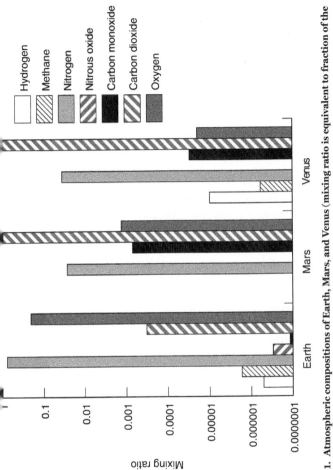

1. Atmospheric compositions of Earth, Mars, and Venus (mixing ratio is equivalent to fraction of the atmosphere).

Laboratory in Pasadena, California, puzzling over what kept the early Earth warm. The puzzle is that stars like our Sun burn steadily brighter over time. When the Earth formed—along with the rest of the Solar System—around 4.5 billion years ago, our Sun was about 30 per cent less luminous than it is today. All else being equal that would have cooled the Earth's surface down by 33°C, meaning that with today's atmospheric composition, the oceans would have been frozen over. Without liquid water at the surface, Earth could not have been the cradle of life. Yet the appearance of sedimentary rocks 3.8 billion years ago showed that material was being weathered from the continents and deposited at the bottom of the sea, and therefore the early Earth did have oceans of liquid water. So something must have kept the early Earth warm.

Sagan suggested that this something could have been a thicker blanket of heat-trapping gases in the atmosphere. His favourite candidate was ammonia, partly because if it was present in the early atmosphere, it could have reacted with other gases to form the basic building blocks of life—amino acids. Now we think that, like Mars and Venus today, the early Earth should have had an atmosphere dominated by carbon dioxide. In the intervening Eons most of that carbon dioxide has been transferred to the Earth's crust. But that just raises a different puzzle—to explain why, as the Sun has got brighter, carbon dioxide was steadily removed, enabling the planet to stay cool.

The Gaia hypothesis

When Lovelock discussed the faint young Sun puzzle with Sagan he had an epiphany: if the Earth's atmosphere is largely a product of life, and its composition has been stable over geological periods of time, perhaps life has been regulating the composition of the atmosphere and thus controlling the Earth's climate. This was to become the Gaia hypothesis—that life and its non-living environment on Earth form a self-regulating system that maintains the Earth's climate and the composition of the atmosphere in a habitable state. Named after the Greek Earth goddess, Lovelock developed the Gaia hypothesis through the late 1960s and early 1970s with the late, great microbiologist Lynn Margulis. It represents the first scientific statement of the Earth as a system that is more than the sum of its parts. Thus for me at least, the Gaia hypothesis marks the start of Earth system science.

Of course there were forerunners who began to think about the Earth as a system and started to recognize the role of life in it. James Hutton, the father of geology in the late 18th century, described the solid Earth as 'not just a machine but also an organized body as it has a regenerative power'. Vladimir Vernadsky in his 1926 book *The Biosphere* argued that life is the key geological force that shapes the Earth. In 1958, the oceanographer Alfred Redfield proposed mechanisms for what he called 'the biological control of chemical factors in the environment'. This is just a tiny sample. Yet no earlier thinker quite saw the global extent and strength of the two-way coupling between life and its planetary environment.

Lovelock and Margulis originally wrote that regulation of the atmosphere was 'by and for' the biota (the sum total of all life on Earth). Although they didn't intend it, this seemed to imply a sort of purposive control of the global environment by unconscious organisms. Such teleological reasoning is out of bounds in science—and thus began a debate over the Gaia hypothesis that continues to this day. In fact, what Lovelock was trying to convey was the idea that a complex system like the Earth can self-regulate automatically, without any conscious foresight or purpose.

Feedback

Lovelock was familiar with systems theory and the branch of it called cybernetics, which studies regulatory systems. A crucial concept in systems theory is feedback. Feedback refers to a chain of cause-and-effect that forms a closed loop (Figure 2). This

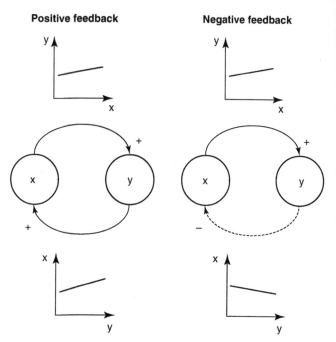

2. Positive and negative feedback. A plus sign on a solid arrow indicates a direct relationship (e.g. increasing x increases y). A minus sign on a dashed arrow indicates an inverse relationship (e.g. increasing y decreases x). An even number (including zero) of inverse relationships in a loop gives a positive feedback, and an odd number gives a negative feedback.

means that information about the past or present state of part of a system can influence its present or future state. Such closed loops of causality can be difficult to get your head around, because we are educated to think 'linearly' in terms of a cause having its effect and that's it—end of story. However, Lovelock realized that in a complex system such as the Earth, there must be a multitude of closed feedback loops that profoundly affect its behaviour.

Feedback comes in two flavours. Positive feedback is an amplifying loop of causal connections—meaning that an initial perturbation to any part of the loop will trigger a response that amplifies the initial change. Negative feedback is a damping loop of causal connections—meaning that an initial perturbation to any part of the loop will trigger a response that damps the initial change. Thus, negative feedback tends to maintain the status quo, whereas positive feedback tends to propel change. 'Positive' and 'negative' are meant here in a mathematical not an emotional sense. 'Positive feedback' is not necessarily a good thing for the Earth system, nor is 'negative feedback' a bad thing. In fact, the mathematical meaning is often the opposite of the emotional meaning.

Lovelock and Margulis postulated that the combination of negative and positive feedback loops in the Earth system produces an overall property of self-regulation—meaning that if something hits the system it tends to bounce back to its original state. This implies that negative feedback has the upper hand, at least near to the starting state of the system. However, the corollary is that if something hits the system too hard it may get propelled into an alternative state, by positive feedback. In other words, self-regulation is not immutable—it can break down.

A key part of Earth system science is identifying the feedback loops in the Earth system and understanding the behaviour they can create. But when Lovelock first had his grand idea of Gaia, he had no idea what the feedback mechanisms that could regulate the climate and the composition of the atmosphere were—and nor did anyone else. Through the 1970s, Lovelock and Margulis began to postulate mechanisms by which the composition of the atmosphere could be regulated, but the long-term stability of the Earth's climate remained a puzzle.

Climate regulation

Then in 1981, James Walker, P. B. Hays, and Jim Kasting proposed a negative feedback mechanism that could counteract the brightening of the Sun and keep the Earth cool. Central to their idea is a process called silicate rock weathering, which removes carbon dioxide from the atmosphere and oceans over geological timescales. This balances the addition of carbon dioxide to the atmosphere and ocean by volcanic and metamorphic processes, which recycle ancient carbon that has been deposited in sediments on the sea floor. In the silicate weathering process, carbon dioxide and rainwater react with silicate rocks, liberating calcium, magnesium, and bicarbonate ions that are washed to the ocean, where they combine to form carbonate rocks. This transfers carbon dioxide from the atmosphere to the Earth's crust.

What Walker and colleagues realized is that silicate rock weathering, like most chemical reactions, occurs faster under warmer conditions. Thus, if something acts to warm the Earth, like the steady brightening of the Sun, this should accelerate the process of silicate rock weathering, and remove more carbon dioxide from the atmosphere. As carbon dioxide is a heat-trapping 'greenhouse' gas this should tend to cool things down again. This is a negative feedback mechanism (Figure 3), and in this case 'negative' feedback is definitely a good thing, because it helps stabilize the Earth's climate.

The silicate weathering negative feedback is not perfect—it cannot completely cancel out the effect of the brightening of the Sun—but it dampens the expected changes in the Earth's temperature. Jim Lovelock, together with Andrew Watson and Mike Whitfield, soon added a biological twist to the feedback mechanism. They noted that plants and their associated soil communities create an acidic weathering environment that dissolves rocks faster, thus speeding up the drawdown of carbon dioxide. This cools the Earth, and because plant productivity responds to changes in carbon dioxide and temperature, it can produce a stronger negative feedback mechanism.

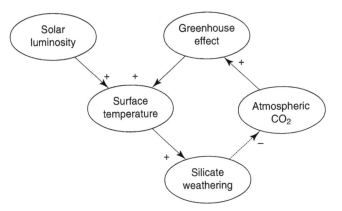

3. The silicate weathering negative feedback, with changing solar luminosity shown as an external forcing.

Snowball Earth

Whilst geochemists were working out what could keep the Earth's climate stable on the longest timescales, early climate modellers were worrying about what might destabilize the climate. In the late 1960s, Mikhail Budyko and William Sellers had independently realized that the Earth's climate could in principle be tipped into a frozen state, where it is covered in ice from equator to poles. This alternative state has become known as 'snowball Earth'—because the planet would appear from the outside like a giant snowball. Remarkably, Budyko and Sellers' models suggested that the snowball state would be stable, just as the present climate state is, because it would absorb far less of the Sun's energy and balance that by giving off less heat radiation, thanks to its lower temperature.

To get from the present climate state to the snowball Earth state involves a particularly strong positive feedback mechanism, known as the 'ice-albedo feedback' (Figure 4). The key idea is that ice and snow are highly reflective (high albedo) to sunlight. Thus,

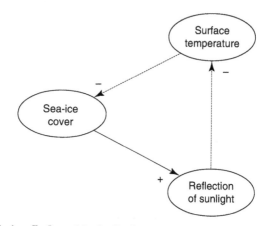

4. The ice-albedo positive feedback.

if something tends to cool the Earth down—like a drop in the carbon dioxide content of the atmosphere—this will cause ice and snow cover to expand, reflecting more sunlight and cooling things down further. In this case 'positive' feedback is definitely not a good thing.

Like all positive feedback mechanisms, the ice-albedo feedback can amplify climate change in either direction; either cooling (with increasing ice cover) or warming (with decreasing ice cover). The feedback operates on today's Earth, with its relatively small ice caps at each pole, tending to amplify climate changes especially near the poles. However, the feedback gets stronger if the planet cools and ice cover increases. This is because the ice is spreading over a sphere into lower latitudes, where there is more incoming solar radiation. As ice reaches to lower latitudes, a given perturbation in temperature can cause a larger incremental change in the area of ice cover, a greater increase in the reflection of sunlight, and a correspondingly larger amplification of the temperature change.

If the ice line reaches roughly 30 degrees in latitude—the tropics—the feedback gets so strong that it 'runs away'. This means that any additional tiny cooling will cause an increase in ice cover and associated cooling that is as large as (or larger than) the initial perturbation. This produces an even larger increase in ice cover, and so on, until the ice snaps shut at the equator creating a snowball Earth.

Such runaway positive feedback is quite rare behaviour. It can only happen when one trip around a positive feedback loop amplifies an initial change by 100 per cent or more. Only a small subset of mechanisms—in the Earth or in any system—can get that strong. Instead the great majority of positive feedback mechanisms are too weak to run away—one trip around the feedback loop amplifies an initial change by much less than 100 per cent—hence the system converges close to where it started.

Escaping a snowball

At the time that Budyko and Sellers formulated their models, hardly anyone thought the Earth had actually been in a snowball state in the past, because it was hard to see how the planet could have escaped from it. However, there is a plausible escape mechanism, first suggested by Joe Kirschvink in 1992. It hinges on shifting the balance of the long-term carbon cycle. Under the arid, freezing climate of a snowball Earth with much of the continents covered in ice sheets, the silicate weathering process that removes carbon dioxide from the atmosphere would be shut off. Yet the input of carbon dioxide to the atmosphere from volcanoes and metamorphic processes would continue—as volcanoes can melt their way through even large ice sheets on the continents. Then with an input but no output, the carbon dioxide concentration of the atmosphere would build up, and up, and up.

As carbon dioxide built up, more of the meagre flux of heat radiation coming off the frozen planet would be trapped and returned to the surface, warming things up. After millions of years, carbon dioxide would eventually reach a concentration sufficient to start melting the ice at the equator and exposing a dark ocean surface. When this happened, the ice-albedo feedback would kick in again, but this time operating in reverse—propelling the melt of ice, once again in a runaway process.

Models suggest that the runaway melt of the snowball Earth would go straight to an ice-free state for the planet. With a vast amount of carbon dioxide in the atmosphere the climate would become very hot (and wet) and the silicate weathering process would go into overdrive. Over the ensuing millions of years, the excess carbon dioxide that had built up in the atmosphere would be removed, cooling the climate down again. Indeed if nothing else changed, the climate could cool to the point that another snowball Earth was triggered and the cycle repeated itself. The resulting oscillation would be typical of the behaviour of a system in which a fast positive feedback—in this case the ice-albedo feedback—interacts with a slow negative feedback process, in this case the silicate weathering feedback. As we will see (in Chapter 4), there is at least one interval in Earth history where multiple snowball Earth events are thought to have occurred.

Global change

It was thinking about the history of our planet, in contrast to its neighbours, that started Earth system science. But by the 1980s another compelling reason to think about the Earth as a system had arisen—the world was waking up to the realization that human activities were changing the present Earth system, on a much shorter timescale. Scientists studying stratospheric ozone depletion and global warming realized that to understand these global changes properly, they had to focus on interactions between physical, chemical, and biological components of the Earth system.

Lovelock's work again played a key role, because in 1971 he was the first to detect the global accumulation of chlorofluorocarbons

(CFCs) in the atmosphere. In 1974, Mario Molina and Sherwood Rowland used a model of atmospheric chemistry to predict that this accumulation of CFCs would catalyse a modest destruction of stratospheric ozone (7 per cent loss over 50–100 years). The reality turned out to be much more dramatic. In 1985, Joe Farman and colleagues published observations of an ozone hole over Antarctica. Ironically, the Total Ozone Mapping Spectrometer satellite instrument, launched in 1979, had been seeing this ozone hole all the time, but a computer algorithm had been rejecting the extreme data as an instrumental error. The opening of the ozone hole triggered frantic scientific research to understand why ozone loss was so extreme. It turned out to hinge on interactions between the polar vortex circulation of the atmosphere, the formation of very cold polar stratospheric clouds, chemical reactions on their surface that liberated chlorine and bromine, and the (already understood) catalytic destruction of ozone by these halogens. By 1987, the Montreal Protocol had been signed by fifty-nine nations, calling for strict limits on CFC emissions.

By the 1980s it was also clear that the concentration of carbon dioxide (CO_2) in the atmosphere was rising, thanks to continuous measurements on Mauna Loa in Hawaii, started by Charles David Keeling in the late 1950s. Anthropogenic sources of CO_2 from fossil fuel burning and land-use change were clearly to blame, but there was a puzzle in that only around half of the CO_2 emissions were accumulating in the atmosphere each year.

To understand what was going on, pioneers like Bert Bolin developed the first models of the global carbon cycle, showing that the ocean and the land were both taking up a fraction of the excess CO_2 that humans were adding. Meanwhile, climate modelling was also maturing. In the late 1960s, Syukuro ('Suki') Manabe and Kirk Bryan had produced the first global model that successfully coupled the circulations of the atmosphere and ocean. Together with Dick Wetherald, Manabe used the model to make the first projections of climate change due to the accumulation of atmospheric CO_2, published in 1975. The observational record of rising global land temperatures was also first compiled during the 1970s, by the Climatic Research Unit at the University of East Anglia. After a series of warm years in the 1980s, climate modeller James Hansen testified to the US Congress in 1988 alerting the world to the problem of global warming.

These two famous examples show how observational scientists and computer modellers began to understand global change in terms of interactions between parts of the Earth system. In the midst of all this activity, NASA brought together a group of scientists to lay out the emerging field of Earth system science. In an influential 1986 report they gave a 'view of the Earth System as a set of interacting processes operating on a wide range of spatial and temporal scales, rather than as a collection of individual components'. The most lasting legacy of the report is a diagram of the interactions between components of the Earth system (Figure 5), which has become known as 'the Bretherton diagram'— after the chair of the committee, Francis Bretherton. What the Bretherton diagram did was put a whole range of existing scientific subjects—and their associated scientific communities— together on the same map. It thus provided the social glue for a wide range of researchers to come together under the integrative banner of 'Earth system science'.

Defining the Earth system

In systems thinking, the first step is usually to identify your system and its boundaries. That means defining what is within the system and what is outside it. The Bretherton diagram (Figure 5) and the accompanying NASA report was one of the first attempts to do just that for the Earth system.

The outer boundary of the Earth system is clear—it is the top of the atmosphere. The Sun is outside the Earth system. It provides our main source of energy, but is not affected by what goes on within the Earth system. Large amounts of energy are exchanged across the top of the atmosphere, but relatively little matter is exchanged. Some hydrogen atoms can escape the Earth's gravity and are lost to space, some meteoritic material comes in (about forty-four tonnes per day on average), but these fluxes of matter are tiny compared to the cycling of matter within the Earth system.

What is less clear is whether and where to put an inner boundary on the Earth system. From the perspective of outer space it is natural to think of the whole of planet Earth as one system. However, the great mass of the inner Earth has its own heat source, fuelled by a combination of radioactive decay and leftover heat from the accretion of the planet. This inner heat source drives convection of the mantle, volcanic activity, and plate tectonics at the surface. Thus it affects the surface Earth system, but it is not affected by it. So in a systems thinking sense it is 'outside' the system, even though it is underneath us (and the flow of liquid iron in the Earth's outer core creates a protective magnetic field around us).

So, where do scientists draw the line between the surface Earth system and the inner Earth? Rather surprisingly, what is part of the Earth system depends on the timescale being considered. If we are concerned with global change over the next century, we exclude the tectonic cycling of the Earth's crust in our models, because that takes place over many millions of years. Indeed we barely need to consider the weathering of the continents and the deposition of sediments in the oceans. We do consider the injection of material by volcanic eruptions, but that is treated as coming from outside the system—which is exactly what the Bretherton diagram shows (Figure 5).

The longer the timescale we look over, the more we need to include in the Earth system. At the extreme, if we are concerned with the mechanisms that have counteracted the steady

5. The 'Bretherton diagram' of fluid and biological Earth processes.

brightening of the Sun over billions of years, we need to consider the creation and movement of the continents, the recycling of carbon deposited in the Earth's crust, and long-term changes in volcanic and tectonic activity. This means that material in the Earth's crust becomes part of the Earth system, and we must recognize that the crust also exchanges material with the Earth's mantle.

All this leads to a rather fuzzy lower boundary to the Earth system. The temptation is to include the whole interior of the planet in the Earth system—and this is exactly what NASA's 1986 report did when considering the longest timescales. Earth scientists also, understandably, tend to keep the whole planet (and therefore the whole of their field) within the bounds of the Earth system. However, for many Earth system scientists, the planet Earth is really comprised of two systems—the surface Earth system that supports life, and the great bulk of the inner Earth underneath. It is the thin layer of a system at the surface of the Earth—and its remarkable properties—that is the subject of this book.

Chapter 2
Recycling

How does today's Earth system support such a flourishing of life? A habitable climate with liquid water is clearly essential, but living organisms also need energy and a host of materials out of which to build their bodies. Energy is in plentiful supply from the Sun, which drives the water cycle and also fuels the biosphere, via photosynthesis. However, the surface Earth system is nearly closed to materials, with only small inputs to the surface from the inner Earth. Thus, to support a flourishing biosphere, all the elements needed by life must be efficiently recycled within the Earth system. This in turn requires energy, to transform materials chemically and to move them physically around the planet. The resulting cycles of matter between the biosphere, atmosphere, ocean, land, and crust are called global biogeochemical cycles—because they involve biological, geological, and chemical processes. This chapter introduces these life-sustaining cycles.

Biogeochemical cycling

The magnitude of recycling that goes on within the Earth system is well illustrated by contrasting the fluxes of gases exchanged at the Earth's surface today with those coming into the surface system from geological processes (Figure 6). The exchange of materials between the Earth's surface and the atmosphere are many orders of magnitude greater than the inputs of materials

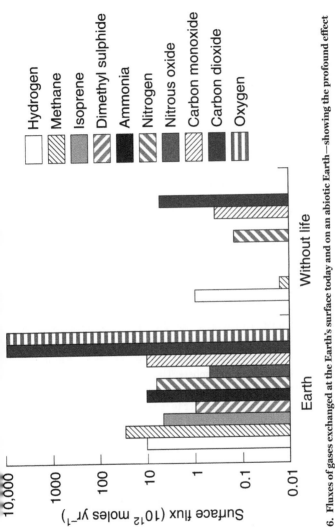

6. Fluxes of gases exchanged at the Earth's surface today and on an abiotic Earth—showing the profound effect of life.

from the solid Earth. This recycling can only be explained by the presence of life on our planet. Indeed several of the key gases being exchanged are uniquely produced by life.

This remarkable gas exchange between the atmosphere and the ocean and land surfaces is just one part of biogeochemical cycling. Elements are also physically transported from the land to the ocean as solids or in solution, carried by the water cycle. The water cycle is the physical circulation of water around the planet, between the ocean (where 97 per cent is stored), atmosphere, ice sheets, glaciers, sea-ice, freshwaters, and groundwater.

The water cycle is intimately tied to the Earth's climate, because it is driven by energy and also carries energy with it. To change the phase of water from solid to liquid or liquid to gas requires energy, which in the climate system comes from the Sun. Equally, when water condenses from gas to liquid or freezes from liquid to solid, energy is released. Solar heating drives evaporation from the ocean. This is responsible for supplying about 90 per cent of the water vapour to the atmosphere, with the other 10 per cent coming from evaporation on the land and freshwater surfaces (and sublimation of ice and snow directly to vapour). Evaporation cools the ocean and land surfaces, and when water vapour condenses to form cloud droplets and rain, or freezes to form snow, this heats the atmosphere. This atmospheric heating in turn drives the upward convection of air masses. The precipitation of rain and snow returns water to the ocean and the land, where it can run off through freshwaters to the ocean. Where precipitated snow remains frozen year-round, an ice sheet can begin to grow.

The water cycle is intimately connected to other biogeochemical cycles (Figure 7). Many compounds are soluble in water, and some react with water. This makes the ocean a key reservoir for several essential elements. It also means that rainwater can scavenge soluble gases and aerosols out of the atmosphere. When rainwater

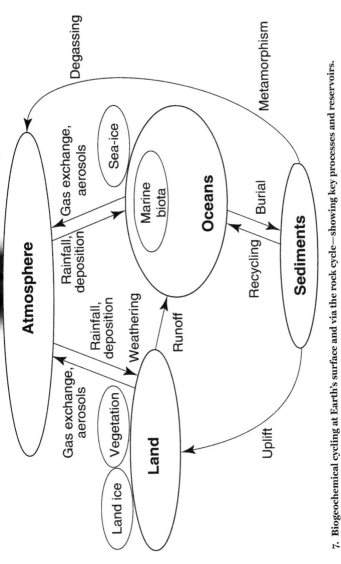

7. Biogeochemical cycling at Earth's surface and via the rock cycle—showing key processes and reservoirs.

hits the land, the resulting solution can chemically weather rocks. Silicate weathering in turn helps keep the climate in a state where water is liquid. Rainfall and glacial scouring also physically erode the land. The particulate and dissolved products of erosion and chemical weathering are transported by freshwaters from the land to the ocean. Once in the ocean, lighter elements can be converted to a gaseous form and returned to the atmosphere, from where a fraction can return to the land (for the heavier elements, gaseous cycling is not an option). Solid materials are deposited from the ocean into its sediments. However, much of what is deposited at the bottom of the ocean is recycled by organisms back to the water column.

Over longer, geological timescales, materials are also recycled via the Earth's crust (Figure 7). Small fluxes of material are temporarily lost from the surface Earth system in new sedimentary rocks forming at the bottom of the sea, but most of this material is ultimately returned to the surface by the rock cycle. Sediments deposited on continental shelves can be later exposed by falls in sea level or uplift of the continental crust. Ocean sediments eventually get subducted at the continental margins, where they are subject to heating and pressurization—metamorphism—releasing the volatile materials they contain back to the atmosphere as gases, sometimes via volcanoes. Metamorphosed rocks are reworked back to the surface by plate tectonics. Together, recycled sedimentary rocks, metamorphic rocks, and igneous rocks (formed from the mantle), provide a fresh supply of materials to the surface Earth system that can be released via the process of chemical weathering.

Each major element important for life has its own global biogeochemical cycle. However, every biogeochemical cycle can be conceptualized as a series of reservoirs (or 'boxes') of material connected by fluxes (or flows) of material between them. Here I will express reservoir sizes in moles (a measure of the number of atoms or molecules they contain) rather than mass (because different elements vary in atomic mass), with fluxes between reservoirs expressed in moles per year. When a biogeochemical cycle is in steady state, the fluxes in and out of each reservoir must be in balance. This allows us to define additional useful quantities. Notably, the amount of material in a reservoir divided by the exchange flux with another reservoir gives the average 'residence time' of material in that reservoir with respect to the chosen process of exchange. For example, there are around 7×10^{16} moles of carbon dioxide (CO_2) in today's atmosphere, and photosynthesis removes around 9×10^{15} moles of CO_2 per year, giving each molecule of CO_2 a residence time of roughly eight years in the atmosphere before it is taken up, somewhere in the world, by photosynthesis.

The oxygen cycle

Photosynthesis is where solar energy enters the biosphere and starts to transform materials chemically. It is apt therefore that the discovery of biogeochemical cycling began with Joseph Priestley's (1772) experiments with plants. Priestley realized that plants exchange materials with the atmosphere as well as the soil. In modern terms, plants acquire their carbon from carbon dioxide in the atmosphere, add electrons derived from water molecules to the carbon, and emit oxygen to the atmosphere as a waste product. The overall reaction, in simplified form, is:

$$CO_2 + H_2O + \text{sunlight} \rightarrow CH_2O + O_2$$

Although plants dominate photosynthesis on land, the first organisms to perform photosynthesis were cyanobacteria, followed by their more complex descendants, algae, and they still dominate primary production in the oceans and freshwaters. In energy terms, global photosynthesis today captures about 130 terrawatts (1 TW = 10^{12} W) of solar energy in chemical form—about half of it in the ocean and about half on land. This removes a massive flux of carbon dioxide from the atmosphere and releases a corresponding amount of molecular oxygen (O_2) gas (Figure 6).

Oxygen is highly reactive stuff and has a strong tendency to rob other elements and compounds of electrons in a process known as 'oxidation'. Materials that have been robbed of electrons, for example by reacting with oxygen, are said to be 'oxidized' (whereas the opposite process of adding electrons is known as 'reduction' and materials with a surfeit of electrons are said to be 'reduced'). Photosynthesis is reversed in the process of aerobic respiration—the oxidation of organic matter with oxygen ($CH_2O + O_2 \rightarrow CO_2 + H_2O$) —releasing the chemical energy that has been captured from sunlight and returning carbon dioxide (oxidized carbon) to the atmosphere. Aerobic respiration is performed by photosynthesizing organisms to fuel their growth, and also by animals, fungi, and a whole range of microbes.

Some of the organic carbon produced in photosynthesis escapes aerobic respiration and reaches places devoid of oxygen—such as ocean sediments or animal guts. There it can be converted back to carbon dioxide by bacteria using nitrate, sulphate, iron oxides, or other oxidized materials. The oxygen in these compounds was originally derived from photosynthesis, so the net result is still to counterbalance photosynthesis, but the reactions yield less energy than aerobic respiration. If oxidized materials run out then a special group of organisms called archaea can convert organic carbon to methane and carbon dioxide—yielding even less energy. The methane ultimately reacts with oxygen in the atmosphere (or other oxidized materials), thus again counterbalancing photosynthesis. All the breakdown pathways for organic carbon together produce a flux of carbon dioxide back to the atmosphere that nearly balances photosynthetic uptake (Figure 6).

The surface recycling system is almost perfect, but a tiny fraction (about 0.1 per cent) of the organic carbon manufactured in photosynthesis escapes recycling and is buried in new sedimentary rocks. This organic carbon burial flux leaves an equivalent amount of oxygen gas behind in the atmosphere. Hence the burial of organic carbon represents the long-term source of oxygen to the atmosphere. It is balanced by the reaction of atmospheric oxygen with ancient organic matter exposed in sedimentary rocks on the continents—a process known as oxidative weathering. There are 3.8×10^{19} moles of molecular oxygen (O_2) in today's atmosphere, and oxidative weathering removes around 1×10^{13} moles of O_2 per year, giving oxygen a residence time of around four million years with respect to removal by oxidative weathering. This makes the oxygen cycle (Figure 8) a geological timescale cycle.

On even longer timescales, some organic carbon and oxygen are exchanged with the Earth's mantle. Oxygen is removed by reacting with reduced volcanic gases coming from the mantle, and ancient organic carbon is added to the mantle when tectonic plates are subducted. The flux of oxidized material to the mantle probably exceeds the flux of reduced material, but the mantle is so massive and well buffered that its oxidation state has not changed much over Earth history. In contrast, the Earth's crust has much more oxygen trapped in rocks in the form of oxidized iron and sulphur, than it has organic carbon. This tells us that there has been a net source of oxygen to the crust over Earth history, which must have come from the loss of hydrogen to space. The full string of reactions is complex, but it starts with the splitting of water in photosynthesis and amounts to losing the hydrogen in water to space and leaving the oxygen behind:

$$2H_2O \rightarrow O_2 + 4H\uparrow_{space}$$

Today only a tiny amount of hydrogen escapes the Earth's atmosphere, making this a tiny oxygen source, but that has not always been the case (as we will see in Chapter 4).

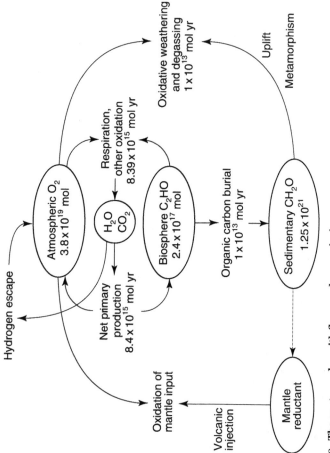

8. The oxygen cycle—with fluxes and reservoir sizes.

The carbon cycle

The oxygen cycle is relatively simple, because the reservoir of oxygen in the atmosphere is so massive that it dwarfs the reservoirs of organic carbon in vegetation, soils, and the ocean. Hence oxygen cannot get used up by the respiration or combustion of organic matter. Even the combustion of all known fossil fuel reserves can only put a small dent in the much larger reservoir of atmospheric oxygen (there are roughly 4×10^{17} moles of fossil fuel carbon, which is only about 1 per cent of the O_2 reservoir).

Carbon dioxide (CO_2), however, is a much scarcer gas than oxygen—there were 750 O_2 molecules for every CO_2 molecule in the atmosphere prior to the Industrial Revolution. Hence the same fluxes can have a much greater proportional effect on CO_2 than O_2. Unlike oxygen, the atmosphere is not the major surface reservoir of carbon. The amount of carbon in global vegetation is comparable to that in the atmosphere and the amount of carbon in soils (including permafrost) is roughly four times that in the atmosphere. Even these reservoirs are dwarfed by the ocean, which stores forty-five times as much carbon as the atmosphere, thanks to the fact that CO_2 reacts with seawater. Therefore, exchange fluxes with the land and ocean have to be considered as potential short-term controls on atmospheric CO_2 (Figure 9).

The exchange of carbon between the atmosphere and the land is largely biological, involving photosynthetic uptake and release by aerobic respiration (and, to a lesser extent, fires). The exchange of carbon between the atmosphere and ocean involves a mixture of chemistry, physics, and biology. CO_2 is continually being exchanged between the surface ocean and the atmosphere. As ocean surface waters physically circulate from the low to the high latitudes they cool down, causing them to take up more CO_2. In some high-latitude regions—the North Atlantic and the Southern

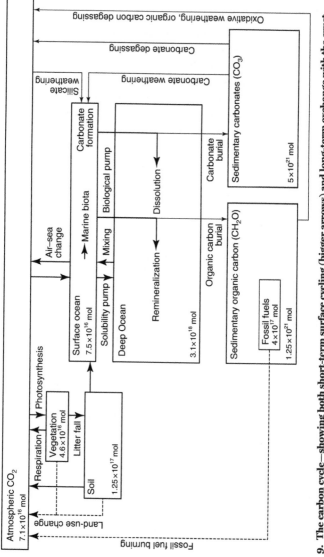

9. The carbon cycle—showing both short-term surface cycling (bigger arrows) and long-term exchange with the crust (smaller arrows), together with major anthropogenic fluxes (dashed arrows).

Ocean today—these surface waters sink to great depth, taking excess CO_2 with them, producing a 'solubility pump' of carbon to the deep ocean. There is also a 'biological pump' of carbon whereby organisms in the surface ocean take up CO_2 and their dead bodies sink (or the excrement of organisms that ate them sinks) transferring carbon to the depths. Both carbon pumps produce an excess of carbon in the deep ocean and a deficit in the atmosphere and the surface ocean.

As well as being made out of organic carbon, some marine organisms precipitate calcium and magnesium carbonates. They include tiny phytoplankton floating in the surface ocean, benthic micro-organisms (that live on the sea floor), and corals. Phytoplankton overproduce carbonate, but as their carbonate shells sink into the deep ocean, they tend to dissolve as the pressure increases. This produces a characteristic 'snowline'—called the 'carbonate compensation depth'—below which no carbonate is preserved in ocean sediments. Above this depth, new carbonate rocks can form, removing carbonate from the surface system. This removal is balanced by the weathering of carbonate rocks on land, which takes up atmospheric carbon dioxide (e.g. $CaCO_3 + CO_2 + H_2O \rightarrow 2HCO_3^- + Ca^{2+}$). When carbonate is precipitated and ultimately preserved in the ocean, this reaction is reversed, releasing one molecule of carbon dioxide for each molecule of carbonate formed (e.g. $2HCO_3^- + Ca^{2+} \rightarrow CaCO_3 + CO_2 + H_2O$). The result is a null cycle with no net effect on atmospheric CO_2, as long as the weathering and deposition of carbonates are in balance.

The long-term carbon cycle involves the exchange of carbon with the crust (Figure 9). The weathering of silicate rocks followed by the formation of carbonate rocks acts as a net removal process for carbon to the Earth's crust (e.g. $CaSiO_3 + CO_2 \rightarrow CaCO_3 + SiO_2$). There are roughly 3.4×10^{18} moles of carbon in the oceans, land, and atmosphere together, and silicate weathering removes around 7×10^{12} moles of carbon per year. Hence the residence time of CO_2

with respect to removal by silicate weathering is around 500,000 years. This carbon returns from the crust when carbonate sediments deposited in the ocean are subducted and 'degassed' by volcanic and metamorphic processes, injecting CO_2 back into the atmosphere. The burial of organic carbon also removes carbon to the crust, which is returned to the Earth's surface by oxidative weathering or degassing (the mirror image of the oxygen cycle). Of the two removal routes to the crust, the flux of carbonate burial is about four times larger than organic carbon burial.

Isotopic constraints

How do Earth system scientists work out such numbers and thus quantify biogeochemical cycling? Of course, they try and estimate fluxes and reservoirs by measuring them directly. But it is not feasible to measure everywhere. Hence models are often used to extrapolate from available measurements to global numbers. When it comes to long-timescale processes involving relatively small fluxes, the error bars can be large. Happily, however, there is an additional data constraint, from the isotopic composition of different reservoirs and fluxes.

Isotopes are especially helpful in reconstructing carbon cycling in the past. Carbon has two stable isotopes—the common ^{12}C and the heavier and rarer ^{13}C (with an extra neutron in the nucleus). Thanks to their different masses, different processes have different preferences for the two different isotopes. For example, photosynthetic uptake of CO_2 (by the enzyme known as RuBisCO) has a strong preference for the lighter ^{12}C over heavier ^{13}C, such that the resulting organic matter is depleted in ^{13}C by about 2.5–3 per cent relative to the atmosphere. This is referred to as 'isotopic fractionation'. It is typically measured on a parts per thousand or 'per mil' scale, relative to some reference material or 'standard' which has a composition defined as 0 per mil. The original standard was a sample of carbonate in the form of the shell of a fossil organism—a belemnite. Fractionations away from the standard are generally small and expressed in a 'delta' notation, e.g. $\delta^{13}C$. Thus, organic carbon produced by photosynthesis has $\delta^{13}C$ of minus 25–30 per mil.

Carbonate rocks average about 0 per mil, but fluctuations away from this value can give valuable clues about past changes in carbon cycling. Notably, if the amount of carbon removed in organic matter changes, then so will the isotopic composition of the ocean, which in turn is recorded in carbonate rocks. For example, if organic carbon burial increases, this will remove more of the lighter ^{12}C from the ocean and make the ocean, and the carbonate rocks formed in it, enriched in ^{13}C. Equally, if there is a decline in organic carbon burial, the ocean and its carbonates will get enriched in ^{12}C.

Remarkably, when we look over Earth history there are fluctuations in the isotopic composition of carbonates, but no net drift up or down. This suggests that there has always been roughly one-fifth of carbon being buried in organic form and the other four-fifths as carbonate rocks. Thus, even on the early Earth, the biosphere was productive enough to support a healthy organic carbon burial flux.

The phosphorus cycle

Productivity and organic carbon burial are determined by the supply of nutrients to the land and ocean. The two most important nutrients for life are phosphorus and nitrogen, and they have very different biogeochemical cycles (Figures 10 and 11). The largest reservoir of nitrogen is in the atmosphere, whereas the heavier phosphorus has no significant gaseous form. Phosphorus thus presents a greater recycling challenge for the biosphere.

All phosphorus enters the surface Earth system from the chemical weathering of rocks on land (Figure 10). Phosphorus is concentrated in rocks in grains or veins of the mineral apatite.

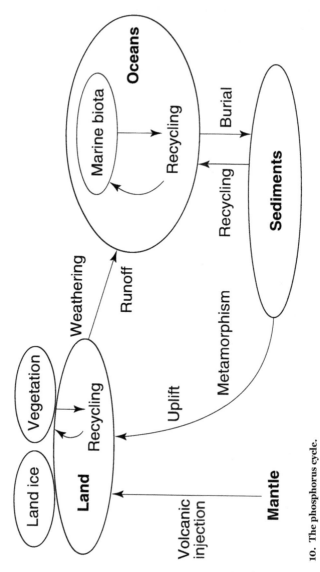

10. The phosphorus cycle.

Natural selection has made plants on land and their fungal partners (called 'mycorrhizae') very effective at acquiring phosphorus from rocks, by manufacturing and secreting a range of organic acids that dissolve apatite. Fungi tunnel through rocks, and when they stumble on apatite they go to work on dissolving it. Once phosphorus is in dissolved form as phosphate it can be taken up directly by plants. However, phosphate can also get adsorbed to the surfaces of minerals such as clays, or react with other elements in soils to form secondary minerals, reducing its availability and putting a premium on phosphorus recycling. Phosphorus in dead organic matter is recycled by bacteria and fungi, including the mycorrhizal fungi that are directly connected to plant roots, limiting the possibility of losses along the way. The average terrestrial ecosystem recycles phosphorus roughly fifty times before it is lost into freshwaters.

The loss of phosphorus from the land is the ocean's gain, providing the key input of this essential nutrient. Phosphorus is stored in the ocean as phosphate dissolved in the water. When phosphate is upwelled from the deep ocean it is taken up by phytoplankton that later die or get eaten. Phosphorus is so highly prized by biology that it is preferentially stripped out of this dead organic matter. This produces a 'microbial loop' of nutrient recycling in the surface ocean that is estimated to boost surface productivity by a factor of three. However, some phosphorus escapes to the deep ocean. Its recycling, via the physical upwelling of water, determines the amount of organic matter that can leave the surface ocean in the 'biological pump' and hence affect the carbon cycle.

Some organic phosphorus reaches the ocean sediments, where again it is preferentially fed on, recycling much of it back to the water column, and helping to support productivity in the surface ocean. However, some phosphorus is trapped in sediments by the formation of new apatite minerals and the adsorption of phosphate on to iron oxide minerals, and some remains in organic matter entering new sedimentary rocks. This removal of phosphorus into the rock cycle balances the weathering of phosphorus from rocks on land. It also dictates the amount of organic carbon that can be buried in the ocean and therefore the long-term source of oxygen to the atmosphere.

The nitrogen cycle

The nitrogen cycle (Figure 11) is controlled by biology. Although there is a large reservoir of nitrogen in the atmosphere, the molecules of nitrogen gas (N_2) are extremely strongly bonded together, making nitrogen unavailable to most organisms. To split N_2 and make nitrogen biologically available requires a remarkable biochemical feat—nitrogen fixation—which uses a lot of energy. In the ocean the dominant nitrogen fixers are cyanobacteria with a direct source of energy from sunlight. On land, various plants form a symbiotic partnership with nitrogen fixing bacteria, making a home for them in root nodules and supplying them with food in return for nitrogen. There are also free-living nitrogen fixers in soil.

Nitrogen is fixed into ammonium, a reduced compound, which when reacted with oxygen yields energy. Nitrifying bacteria live off this energy, converting ammonium to nitrite and then nitrate (NO_3^-) in the process of nitrification. In our oxygenated world, nitrate is a relatively stable compound which forms the major reservoir of biologically available nitrogen in the ocean and soils. However, nitrate can be used to oxidize organic matter yielding energy and converting the nitrate ultimately back to nitrogen gas (N_2), in the process of denitrification. Denitrifying bacteria live off this energy source, and nitrous oxide (N_2O) gas is often also released along the way. Denitrification is favoured once oxygen falls to a low concentration, so it tends to occur in waterlogged, anoxic soils and in 'oxygen minimum zones' at intermediate depth in the ocean (which are created by the aerobic respiration of sinking organic matter).

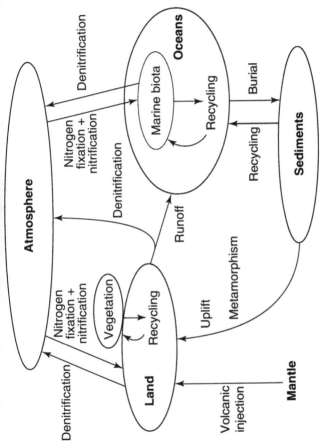

11. The nitrogen cycle.

Nitrogen fixation and denitrification form the major input and output fluxes of nitrogen to both the land and the ocean, but there is also recycling of nitrogen within ecosystems. Photosynthesizers (plants on land, phytoplankton in the ocean) take up nitrate (or ammonium in the case of some phytoplankton) and assimilate it into organic matter. This nitrogen can then be recovered from dead organic matter by bacteria, and also fungi on land, that convert it to ammonium in a process known as ammonification (or remineralization). The ammonium can then be nitrified to nitrate and taken up again by photosynthesizers. In the ocean, the remineralization of nitrogen from organic matter takes place over a range of depths as dead organic matter sinks through the water column. Like phosphorus, some nitrogen is recycled fast in the well-mixed, sunlit surface layers of the ocean. The rest is recycled slowly in dark, deeper waters and has to await the physical upwelling of those waters before it can be taken up again by phytoplankton.

Land ecosystems leak some nitrogen to the ocean in organic matter and as dissolved nitrate. Much of this nitrogen gets denitrified in estuaries and coastal shelf sea sediments. Some of the nitrogen fixed in the ocean reaches the sediments in organic matter. Again, much of it is denitrified in the sediments. However, there is a small loss of nitrogen into new sedimentary rocks. There are 1.4×10^{20} moles of N_2 in the atmosphere and nitrogen is buried in the crust at a rate of around 3×10^{11} moles per year, giving nitrogen a residence time of around 500 million years in the atmosphere with respect to removal to the crust. Whilst this is a long time, the removal of nitrogen to the crust must nevertheless have been roughly balanced over Earth history by inputs of nitrogen from the mantle and recycling via rock weathering.

Earth's metabolism

The global biogeochemical cycling of materials, fuelled by solar energy, has transformed the Earth system. This transformation of energy and cycling of materials by the biosphere can be seen as the 'metabolism' of the Earth system. It is fundamental to the remarkable productivity of the Earth's biosphere, just as an individual organism's metabolism is essential to its healthy existence. It has made the Earth fundamentally different from its state before life and from its planetary neighbours, Mars and Venus. Through cycling the materials it needs, the Earth's biosphere has bootstrapped itself into a much more productive state. In Chapter 3 we consider how the biogeochemical cycles are self-regulated and how they are coupled to the Earth's climate.

Chapter 3
Regulation

The Earth system has maintained habitable conditions for life over geological periods of time. These conditions include an equable global temperature, enough atmospheric carbon dioxide to fuel photosynthesis, and sufficient nutrients for growth. Furthermore, for at least the past 370 million years, there has been enough atmospheric oxygen to support complex, mobile animal life, but not so much that wildfires decimated vegetation. This chapter introduces the ways in which these 'master variables' of the Earth system are regulated and how scientists study this regulation.

Basic concepts

Negative feedback is at the heart of all regulatory mechanisms—in general terms it is a closed loop of causal connections that tends to damp perturbations to any part of the loop. When thinking about the regulation of materials—like nutrients in the ocean, or carbon dioxide or oxygen in the atmosphere—we need to link the concept of negative feedback (introduced in Chapter 1) to the concept of reservoirs and fluxes (introduced in Chapter 2). In basic terms, to regulate the size of a reservoir of material, negative feedback can operate either on the input flux to that reservoir or the output flux from it. For example, as the size of a reservoir increases, negative feedback could increase the output flux from that reservoir, thus stabilizing it.

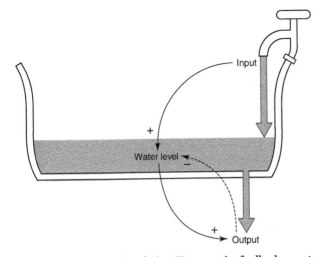

12. The bath metaphor of regulation. Here negative feedback operates on the output from a reservoir.

One way to visualize this is with the metaphor of a bath (Figure 12). It has an input (from the tap) and an output (down the plughole). Turn the tap on with the plughole open, and the bath water should find a stable level which will depend on how much you turned the tap on, because the output flux down the plughole tends to increase with the volume of water in the bath—providing a negative feedback. Imagine if the area of the plughole increased as the volume of water in the bath increased—that would give an even better negative feedback. But what if the water was piped away at a constant rate regardless of how much is in the bath? Then there would be no feedback control on the output from the bath. For this system to be stable there would then need to be feedback on the input to the bath that counteracts changes in the water level by adjusting the tap—turning it down as the water level rises or turning it up as the water level drops.

The water in the bath metaphor could represent a reservoir of any material—it doesn't have to be water or even liquid. As we will see, real cases of regulation in the Earth system sometimes involve negative feedback on the input to a reservoir, and sometimes on the output from a reservoir.

Biogeochemical models

To try to understand what regulates nutrients, oxygen, and carbon dioxide over geological timescales, Earth system scientists build models. These long-term biogeochemical models represent key reservoirs as a series of boxes with fluxes between them. The task of the modeller is to identify where and how the fluxes in the model depend on the reservoir sizes, thus producing feedbacks—as in the bath metaphor. Often these feedbacks operate via intermediate variables such as temperature, which are not reservoirs of material in themselves, but are affected by them.

In this kind of research, the model acts as a tool to help understanding. It represents the modeller's mechanistic hypotheses for how they think the world works and it makes predictions from them, which can be tested against available data. Typically the modeller experiments by putting some new process or feedback connection into the model (representing a hypothesis), and looks at how it affects the model's predictions with respect to some data (observations) that they are trying to explain. If the new process moves the results away from the data the hypothesis is likely to be falsified; if it moves the results towards the data the hypothesis survives. (In reality, things are sometimes more subtle than this because a model of a complex feedback system can, like the real world, display surprising behaviour and several simultaneous adjustments may bring it back towards the data.)

I developed such a model as a PhD student with Andrew Watson, to address the interconnected questions: what regulates the nutrient balance of the ocean and the oxygen content of the atmosphere? Later we extended the model—with another student, Noam Bergman—to consider: what regulates the carbon dioxide content of the atmosphere and the climate on geological timescales? The resulting model, called COPSE, built on pioneering work by Bob Berner, who has made a series of models to understand variations in the long-term carbon and oxygen cycles. The target for all these models has been the last 542 million years, known as the Phanerozoic Eon—literally, the era of visible life, starting with animals and witnessing the rise of land plants. The next sections outline the puzzles we were trying to solve and what we learned with the help of the model.

Nutrient regulation

There is a remarkable correspondence between the ratio in which the essential nutrients nitrogen and phosphorus are found in ocean water and the ratio in which they are required by marine organisms. This was first highlighted by the oceanographer Alfred Redfield in 1934 and the average N:P ratio of marine organisms is known as the 'Redfield ratio' in his honour. Traditionally phytoplankton have N:P = 16 whereas waters upwelling from the deep ocean have close to N:P = 15. So from the average organisms' point of view there is a slight deficit of nitrogen relative to phosphorus in the ocean. The 'Redfield puzzle' is to explain what gives rise to the correspondence between ocean composition and organism composition. Could it have happened by chance? Have the organisms simply adapted to the environmental conditions? Or has life somehow regulated the ocean composition to match its own requirements? Redfield argued for the last option.

Key to Redfield's feedback mechanism is the activities of nitrogen fixing organisms. Fixing nitrogen comes at a serious energy cost (in splitting the triple bond of N_2), which means that nitrogen fixers tend to be outcompeted by non-fixers whenever nitrogen is available. Thus, when deep waters with N:P < 16 are physically upwelled to the surface ocean the nitrogen they contain gets used up by other phytoplankton, but when nitrogen runs out, on average, some phosphorus will remain. Nitrogen fixing organisms then have the opportunity to grow using this remaining phosphorus and fixing nitrogen directly from the atmosphere. The activity of nitrogen fixers adds fixed nitrogen to the ocean, so they limit their own spread. However, fixed nitrogen is continually being removed by the activities of denitrifying organisms that thrive wherever oxygen runs out at depth in the ocean. This allows some nitrogen fixers to persist as a small fraction of the surface community.

The result is a negative feedback mechanism operating on the input of nitrogen to the ocean and keeping it in balance with the outputs. This feedback enables the nitrogen content of the ocean to track variations in the phosphorus content, which can be driven by, for example, fluctuations in weathering (Figure 13). If the input of phosphorus to the ocean increases then nitrogen fixation will increase, increasing the nitrogen content of the ocean. If phosphorus input drops then nitrogen fixation will decrease, leaving denitrification to reduce the nitrogen content of the ocean. Equally, changes in the removal of phosphorus from the ocean will trigger counteracting changes in nitrogen fixation and the nitrogen content of the ocean. Thus, even though nitrogen is usually the first nutrient to run out in the surface ocean, because nitrogen can track variations in phosphorus, phosphorus is seen as the 'ultimate' limiting nutrient on geologic timescales.

The phosphorus content of the ocean is also regulated, but by negative feedback operating on its output from the ocean—because the input of phosphorus down rivers cannot be controlled by processes in the ocean. If the input of phosphorus to the ocean and hence its concentration increases, this increases nitrogen, productivity, and the removal of phosphorus into marine sediments. Equally, if the input of phosphorus declines, this reduces its concentration in the ocean, lowering surface productivity and suppressing phosphorus removal to sediments. The resulting negative feedback is not perfect, but it buffers fluctuations in the

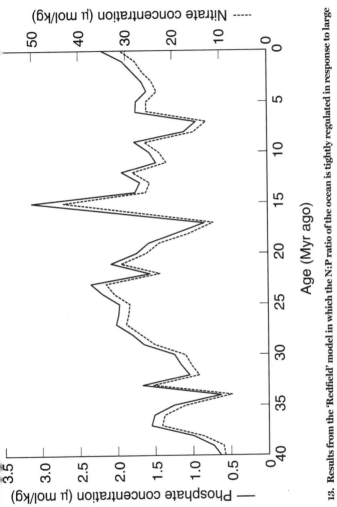

13. Results from the 'Redfield' model in which the N:P ratio of the ocean is tightly regulated in response to large imposed fluctuations in P input.

nutrient concentration of the ocean, making them smaller than driving fluctuations in phosphorus input to the ocean.

There is an intimate link between nutrient regulation and atmospheric oxygen regulation, because nutrient levels and marine productivity determine the source of oxygen via organic carbon burial. However, ocean nutrients are regulated on a much shorter timescale than atmospheric oxygen because their residence times are much shorter—about 2,000 years for nitrogen and 20,000 years for phosphorus.

Oxygen regulation

The residence time of oxygen in the atmosphere with respect to removal into the Earth's crust is around four million years (see Chapter 2). This may sound like a long time, but it is far shorter than the 550 million years or so over which there have been oxygen-breathing animals on the planet. It is also far shorter than the 370 million years over which there have been forests, which are vulnerable to increases in oxygen that increase the frequency and ferocity of fires. Thus, remarkably, the amount of atmospheric oxygen has remained within habitable bounds for complex animal and plant life despite all of the oxygen molecules having been replaced over a hundred times.

In fact, the stability of oxygen has been even more remarkable since the spread of forests over the planet. Combustion experiments show that fires only become self-sustaining in natural fuels when oxygen reaches around 17 per cent of the atmosphere. Yet for the last 370 million years there is a nearly continuous record of fossil charcoal, indicating that oxygen has never dropped below this level. At the same time, oxygen has never risen too high for fires to have prevented the slow regeneration of forests. The ease of combustion increases non-linearly with oxygen concentration, such that above 25–30 per cent oxygen (depending on the wetness of fuel) it is hard to see how forests could have survived. Thus oxygen has remained within 17–30 per cent of the atmosphere for at least the last 370 million years. The question is: what negative feedback mechanisms can explain this remarkable stability?

In principle there are two places where stabilizing feedback could occur—on the long-term source of oxygen from organic carbon burial or on the long-term sink of oxygen from oxidative weathering. However, in today's oxygen-rich world, most of the ancient organic carbon exposed by the uplift of rocks on the continents gets oxidized. Therefore there is little scope for small variations in atmospheric oxygen concentration to affect the removal flux of oxygen via oxidative weathering. Instead there must be mechanisms by which the source flux of oxygen from organic carbon burial is sensitive to variations in oxygen concentration.

In today's world roughly half of the organic carbon that gets buried comes from primary production in the ocean and half from primary production on land, but it is nearly all buried in ocean sediments. So it is natural to look for oxygen regulation mechanisms in the ocean. When atmospheric oxygen levels drop, so does the concentration of oxygen in the ocean, and when the ocean gets more anoxic (devoid of oxygen) more organic carbon gets preserved in sediments, acting as a negative feedback that tends to increase atmospheric oxygen. This mechanism turns out to involve the phosphorus cycle—under anoxic conditions, more phosphorus is recycled from ocean sediments back to the water column, and this fuels more productivity in the surface ocean, increasing the supply of organic carbon to depth and its burial.

This ocean-based regulatory mechanism helps buffer against declines in atmospheric oxygen. However, if oxygen rises above present levels, the whole ocean becomes oxygenated and the feedback switches off. Therefore we have to look to the land for a sensitive mechanism that can counteract rises in oxygen. Fires and vegetation are obvious candidates to be involved. As fires suppress vegetation this reduces the supply of terrestrial organic material for burial, but tends to transfer phosphorus from the land to the ocean, fuelling productivity there. However, marine organic material has a much lower carbon-to-phosphorus ratio than terrestrial plant matter, which means that the same global phosphorus supply supports less organic carbon burial, weakening the oxygen source. In addition, when fires suppress forests this reduces their effect on rock weathering and therefore phosphorus input, reducing organic carbon burial and oxygen production.

Including these feedback mechanisms in a biogeochemical model, they can explain the long-term stability of atmospheric oxygen (Figure 14). The model predicts that oxygen remained within 17–30 per cent of the atmosphere over the last 350 million years, consistent with the record of charcoal and forests. It predicts 5–10 per cent atmospheric oxygen prior to the rise of plants, due to a weaker source of phosphorus (slower rock weathering) and a lower carbon-to-phosphorus burial ratio. However, this is still sufficient oxygen to explain the existence of early animals. Prior to plants, with more widespread anoxia in the ocean, the model suggests that oxygen was stabilized by ocean-based negative feedback mechanisms.

The oxygen cycle is intimately connected to the long-term carbon cycle. The dominant removal process for carbon dioxide (CO_2) through silicate rock weathering is a major source of weathered phosphorus, which in turn controls organic carbon burial and hence the source of oxygen. The burial of organic carbon is also the second most important sink of atmospheric CO_2.

Long-term carbon dioxide regulation

On the longest geological timescales, the concentration of atmospheric CO_2 has been regulated by the dependence of silicate weathering on atmospheric CO_2 and global temperature. We met this important negative feedback mechanism in Chapter 1 (Figure 3). To recap, the rate of silicate weathering increases with

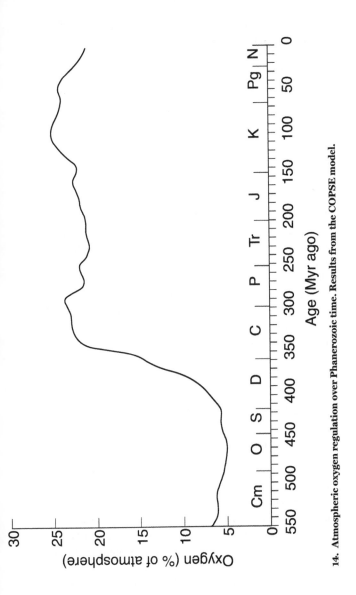

14. Atmospheric oxygen regulation over Phanerozoic time. Results from the COPSE model.

increasing CO_2 and temperature. Thus, if something tends to increase CO_2 or temperature it is counteracted by increased CO_2 removal by silicate weathering. Equally, if something tends to decrease CO_2 or temperature, less CO_2 is removed by silicate weathering. In today's world, this crucial feedback is largely under the control of land plants and their associated mycorrhizal fungi. Plants are sensitive to variations in CO_2 and temperature, and together with their fungal partners they greatly amplify weathering rates (as discussed in Chapter 2). The result is a stronger negative feedback mechanism than in the absence of land life.

Models of the long-term carbon cycle include this feedback together with multiple geological and biological drivers of CO_2 variation. For example, the source of atmospheric CO_2 from volcanic and metamorphic processes is estimated to have fluctuated with changes in plate tectonics, and the sink of CO_2 has fluctuated with varying uplift of the continents and the eruption of easily weathered basalts on to the land. Still, the most pronounced change in atmospheric CO_2 over Phanerozoic time was due to plants colonizing the land. This started around 470 million years ago and escalated with the first forests 370 million years ago. The resulting acceleration of silicate weathering is estimated to have lowered the concentration of atmospheric CO_2 by an order of magnitude (Figure 15) and cooled the planet into a series of ice ages in the Carboniferous and Permian Periods.

Shorter-term carbon dioxide regulation

The silicate weathering feedback tends to stabilize atmospheric CO_2 and global temperature over a timescale of hundreds of thousands of years. However, geologic perturbations such as massive volcanic eruptions or the sudden metamorphism of organic-rich sediments can occasionally add excess CO_2 to the atmosphere much faster than this, overwhelming the negative feedback. Human activities are also now adding CO_2 to the atmosphere at an unprecedented rate. Happily there are a series

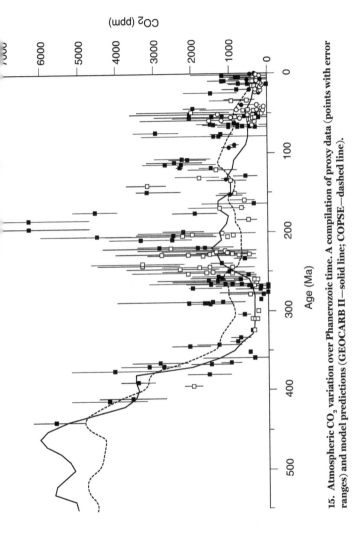

15. Atmospheric CO₂ variation over Phanerozoic time. A compilation of proxy data (points with error ranges) and model predictions (GEOCARB II—solid line; COPSE—dashed line).

of other negative feedback mechanisms that can regulate atmospheric CO_2 over shorter timescales.

On the timescale of years to centuries, excess CO_2 added to the atmosphere starts to get taken up by the ocean and by the land biosphere. After about a thousand years a temporary balance is found, with the added CO_2 apportioned between the oceans, atmosphere, and land surface. At least 15 per cent of the CO_2 added remains in the atmosphere on the millennial timescale, with that fraction increasing the more CO_2 is added in the first place—because the CO_2 that has dissolved in the ocean acidifies it and reduces its capacity to store carbon.

This in turn triggers a negative feedback mechanism known as 'carbonate compensation' that tends to remove more of the added CO_2 from that atmosphere over a timescale of roughly 10,000 years. In this mechanism, acidified ocean waters tend to dissolve carbonate sediments in contact with them, raising the carbonate compensation depth (CCD). Importantly, carbonate sediments contain alkalinity in a 2:1 ratio with carbon, and it is the amount of alkalinity in the ocean that dictates its storage capacity for carbon. Thus, the dissolution of carbonate sediments adds more alkalinity than carbon to the ocean allowing it to take up more CO_2 from the atmosphere.

Meanwhile, the excess CO_2 in the atmosphere tends to accelerate the weathering of carbonate rocks on land—by increasing temperature and acidifying rainwater. This increases the supply of alkalinity to the ocean, further increasing its storage capacity for CO_2, again providing a negative feedback on a roughly 10,000-year timescale. Ultimately this resupply of alkalinity allows carbonate sediments to be redeposited at depth in the ocean—the CCD gets deeper again and the carbonate cycle refinds a balance. After all this has played out, a small fraction of the originally added CO_2 remains in the atmosphere and is removed over hundreds of thousands of years by the silicate weathering feedback.

A historical example

Can we see any examples of carbon dioxide and climate regulation operating in the geological record? Large natural perturbations to the carbon cycle are fairly rare but nevertheless there are several in the geological record, one of the more recent ones being 55.8 million years ago at the boundary of the Palaeocene and Eocene Epochs. Known as the Palaeocene-Eocene Thermal Maximum—or PETM for short—this striking warm event offers some important clues as to where we might be sending the climate in the future, and how long it will take to recover.

No one is absolutely sure what caused the PETM, but we know there was a large injection of thousands of billions of tonnes of carbon into the atmosphere, probably triggered by a volcanic intrusion into ancient fossil fuel reserves, and supplemented by the destabilization of frozen methane hydrates under ocean sediments. The carbon appears to have been injected in two pulses 20,000 years apart. Global temperatures rose roughly 5°C over 20,000 years and remained high for around 100,000 years. Acidification of the ocean led to widespread dissolution of carbonate sediments, with the CCD rising by up to 2 km. It took around 200,000 years for the carbon cycle and climate to recover fully.

The slow recovery from the PETM is consistent with the timescale of the silicate weathering feedback. It should caution us that even though there are multiple regulatory feedbacks in the carbon cycle, they can be overwhelmed. Hence human fossil fuel burning activities can be expected to leave a similarly long climate legacy.

Biogeochemical climate feedbacks

Whilst carbon dioxide has played a fundamental role in climate regulation over Earth history, there are other key players as well.

In particular, changes in the albedo or reflectivity of the Earth can have a large leverage on Earth's temperature.

Clouds are vital to determining the Earth's albedo. Whilst they seem to us wholly physical things, clouds can be affected by biology, because cloud water needs something to condense on. A variety of biological gases produce aerosol particles that in turn form nucleation sites on which water vapour can condense to form clouds. In particular, marine phytoplankton release a gas called dimethyl sulphide—or DMS for short—which is the major source of cloud condensation nuclei (CCN) over remote, unpolluted parts of the ocean today (and prior to human industrial pollution was even more important as a global source of condensation nuclei). Increasing the number of CCN in a cloud distributes the same amount of water over a larger number of smaller droplets, which makes the cloud whiter—meaning it reflects more sunlight. This biological production of CCN cools the Earth by several degrees.

The realization that DMS is a major source of CCN led Bob Charlson, Jim Lovelock, Andy Andreae, and Steve Warren famously to propose a climate feedback—known as the CLAW hypothesis after the authors' initials (Figure 16). They argued that if something acts to increase temperature or sunlight incident on the surface ocean, this should increase the biological production of DMS, leading to more reflective clouds that reflect sunlight back to space, cooling things down again. In an unpolluted, pre-human world this negative feedback may have been an important short-term climate regulator. However, if temperature gets too high, the surface ocean begins to stratify, restricting the supply of nutrients from below and thus limiting biological production and the production of DMS. This switches the sign of the temperature feedback from negative to positive causing it to amplify climate change.

In fact, several other biogeochemical feedbacks on climate are positive rather than negative. Like most biological processes, the

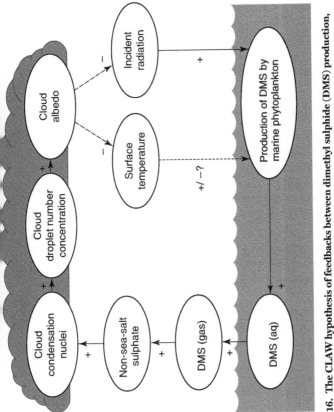

16. The CLAW hypothesis of feedbacks between dimethyl sulphide (DMS) production, cloud albedo, and climate.

production of the biological 'greenhouse' gases carbon dioxide (CO_2), methane (CH_4), and nitrous oxide (N_2O) all increase with temperature. Thus if something acts to increase temperature, the source of these gases will tend to increase, increasing the temperature further. These positive feedbacks act on a physical climate system that is already in a state of overall positive feedback due to physical mechanisms. In particular, the most important greenhouse gas of all, water vapour, increases in concentration with warming caused by the other greenhouse gases, thus amplifying their effects.

How good is Earth's climate regulation?

The examples touched on suggest that whilst the Earth's climate system contains at least one long-term stabilizing mechanism, it also contains a mixture of shorter-term stabilizing and destabilizing feedbacks. The long persistence of life suggests the climate is regulated between broad bounds, but the snowball Earth hypothesis (introduced in Chapter 1) suggests that climate regulation can sometimes break down quite catastrophically. A key question then is: how stable is the current climate system? The record of recent past climate changes provides some useful clues to an answer.

Over the last forty million years the Earth has been cooling such that around 2.5 million years ago, Northern Hemisphere ice age cycles began. Initially these had a period of around 41,000 years (linked to periodic fluctuations in the tilt of the Earth's axis), but in the last million years the ice ages have become longer and deeper with a roughly 100,000-year period (Figure 17). These recent ice age cycles provide a beautiful example of the Earth functioning as a whole system, which is apparently exquisitely sensitive to subtle changes in the Earth's orbit, with feedbacks within the system dominating its behaviour. The climate and the carbon cycle have fluctuated in synchrony as revealed by the ice core record, with times of warming being times of rising CO_2, CH_4,

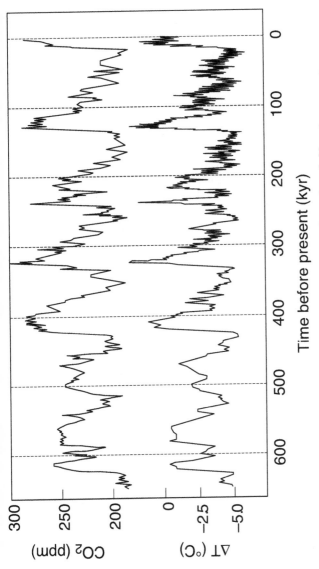

17. The Antarctic ice core record of atmospheric CO_2 and temperature change (rescaled here to represent approximately global temperature change).

and N_2O (and cooling the converse). At the end of each ice age, the positive feedback between atmospheric CO_2 and temperature was strong enough that the climate may temporarily have gone into a 'runaway' state, abruptly shifting the state of the whole planet from glacial to interglacial conditions.

This sense of instability is reinforced when we look within the last ice age at shorter-term climate fluctuations. There were repeated, incredibly rapid climate changes that were at least hemispheric in the extent of their impacts. As the last ice age ended, our record of these abrupt climate changes comes into sharper focus, revealing that warming of up to 10°C in Greenland has occurred within less than a decade. This reinforces the idea that the present climate system is unusually unstable—at least on relatively short timescales—providing an important backdrop for thinking about our own planet-changing activities as a species (see Chapters 5–7).

Changing stability

Key variables in the Earth system—ocean nitrogen and phosphorus, and atmospheric oxygen and carbon dioxide—are regulated by negative feedback mechanisms involving living organisms. These feedbacks have maintained fairly stable conditions on Earth for hundreds of millions of years. Stability in this systems sense does not mean constancy—changing biological and geological drivers have produced long-term changes in the Earth system. In particular, the rise of land plants caused a rise in atmospheric oxygen and a fall in carbon dioxide and temperature. Still, these changes were a lot smaller than they would have been in the absence of negative feedbacks. Whilst the Earth's climate has been stabilized on geological timescales, it shows evidence of instability on shorter timescales, especially approaching the present time. Having introduced how the Earth system with complex life self-regulates, Chapter 4 considers the fundamental changes that created it.

Chapter 4
Revolutions

How did today's Earth system become so radically different from those of our planetary neighbours, Mars and Venus? This is a big question, but the presence of life is clearly a big part of the answer. The past few decades have witnessed remarkable scientific progress in our understanding of the development of the Earth as a system and how it is coupled to the evolution of life. Earth system scientists now think in terms of the coupled evolution of life and the planet, recognizing that the evolution of life has shaped the planet, changes in the planetary environment have shaped life, and together they can be viewed as one process. When we look at this 'co-evolution' over Earth history, a relatively few revolutionary changes leap out, in which the Earth system was radically transformed. Each of these revolutionary changes depended on the previous one, and without them we would not be here. This chapter delves deeper into Earth history to introduce those revolutions.

The evidence

To understand the science behind this remarkable story we need to appreciate the hard-won evidence behind it. The key source of evidence is ancient rocks still exposed at the Earth's surface. Particularly valuable are sedimentary rocks that were deposited from the ancient oceans. Occasionally soils from the ancient land surface are also preserved. Both can give clues about the past composition of the oceans and atmosphere. Many of those clues are written in the proportions of elements contained in the sediments and their isotopic composition. Ancient sediments may also contain fossils, which for much of Earth history are rare and only visible through a microscope. Very occasionally sediments contain organic-rich oils which can yield 'molecular fossils' called biomarkers. These special organic compounds are made by only a subset of organisms, therefore revealing the presence of those organisms.

The history of life is also written in the genetic code of organisms alive today. When compared, the degree of genetic difference between organisms can be used to reconstruct a 'phylogenetic tree', which shows the ordering in which different lineages split from a common ancestor. If we are confident in the phylogenetic tree and we know something about the rate of genetic mutations on different branches of the tree, then it can also be turned into a 'molecular clock'. This uses the degree of genetic difference between organisms—on non-coding parts of the genome not subject to selection—to determine how long ago their lineages split from a common ancestor: i.e. the length (in time) of different branches of the phylogenetic tree. Molecular clocks are calibrated against the fossil record in relatively recent time, and can then be used to extrapolate further back in time, where the fossil record is sparse. Early molecular clock estimates had such large error bars that they told us very little about the timing of origin of ancient lineages of life. However, refinements of the method are producing more precise timing estimates that also appear to be more accurate when compared to the sparse fossil record.

Deep time

To get a handle on the history of the Earth system we need to grapple with deep time—events happening over the course of billions of years. This means shifting our perception of time from that of our day-to-day lives to that of the Earth system's geological processes. This can be rather disorientating—as one of the first people to be confronted with it (by his friend James Hutton), John Playfair remarked: 'the mind seemed to grow giddy by looking so far into the abyss of time'. To help us along the way I will tell the story chronologically, ordering events on a timeline (Figure 18). Earth scientists use the shorthand 'Ga' for billions of years ago, and 'Ma' for millions of years ago. As a first cut they divide Earth history into four Eons; the Hadean, Archean, Proterozoic, and Phanerozoic Eons.

Crucial to improved understanding of Earth history has been the ability to date rocks accurately and thus establish the timing of key events. This enables scientists to arrange the evidence we have in a chronological sequence and begin to make inferences about causality. Radiometric dating uses the radioactive decay of different long-lived isotopes. The most widely used technique is uranium-lead dating of tiny grains of the common mineral zircon found in ancient rocks. This technique can take advantage of the fact that there are two different long-lived isotopes of uranium that decay to two different isotopes of lead. This enables cross-checking of the dating and yields remarkably precise age estimates.

Origin of the Earth

Our timeline starts with the formation of the Solar System. This is dated from the oldest meteoritic material at 4.567 billion years ago. The Earth and the other planets are younger than this, because they had to form from the gravitational collisions and accumulation of material spinning around the early Sun—in a process called accretion. During the accretion of the Earth there were some truly massive collisions, the last of which is thought to have formed the Moon 4.470 billion years ago. An object named Theia (after the mother of Selene, the goddess of the Moon) collided with the Earth, ejecting a mass of material that accreted

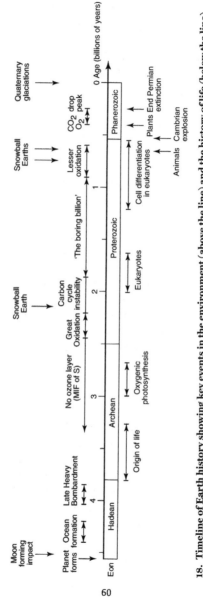

18. Timeline of Earth history showing key events in the environment (above the line) and the history of life (below the line).

to form the Moon. We can be fairly sure of this because the Moon is less dense than the Earth, indicating that it lacks an iron-rich core.

At this point the Earth was still forming, but the gas giants Jupiter and Saturn had finished accreting. Their gravitational pull disrupted the band of asteroids between Mars and Jupiter, sending some of them off on elliptical orbits that crossed the inner Solar System. Crucially, this brought water and other volatile substances, including nitrogen and carbon dioxide, to the early Earth (as well as to Mars and Venus). Remarkably, some tiny bits of the Earth's crust from this time are still present at the surface today, in the form of grains of zircon. The oldest is 4.374 billion years old and was originally part of a granite rock, indicating that the continental crust had started to form in the first 100 million years of the planet's history. The isotopic composition of oxygen in the zircon also suggests that oceans of liquid water were present on the Earth at the time.

The onslaught from outer space was not over though. The Earth and all the inner Solar System suffered a 'Late Heavy Bombardment' by asteroids. This was revealed by dating of impact-melted rocks brought back from the Moon by the Apollo missions, which showed a clustering of ages in the range 4.1–3.8 billion years ago. Computer simulations suggest that the Late Heavy Bombardment could have been caused by a resonance between the orbits of Jupiter and Saturn, which deflected asteroids on elliptical orbits into the inner Solar System. Some of the impacts in the bombardment were large enough to have evaporated the early oceans and thus temporarily rendered the planet uninhabitable (or nearly so).

Origin of life

The most fundamental change in the history of the Earth system was the origin of life. The first tentative evidence for life on Earth comes remarkably soon after the end of the Late Heavy Bombardment, around 3.8 billion years ago. It takes the form of small particles of graphite—made from organic carbon—in some of the oldest sedimentary rocks. The first putative microscopic fossils of life are about 3.5 billion years old. They look like bacteria, but not everyone is convinced the fossil structures were made by biology. The first widely agreed evidence for life is 3.26 billion-year-old microfossils apparently catching bacteria in the act of cell division.

So, what fuelled the early biosphere? The earliest life forms may have consumed compounds in their environment that could be reacted to release chemical energy. However, supplies of chemical energy are generally small, except in unusual environments such as deep sea vents near mid-ocean ridges. Shortage of chemical energy on a global scale would thus have restricted the productivity of life. One possibility is that early archaea consumed hydrogen from the atmosphere and carbon dioxide to make methane, but such a methanogen-based biosphere would have been restricted to around a thousandth of the productivity of the modern marine biosphere.

A more productive global biosphere would have arisen when early life began to harness the most abundant energy source on the planet—sunlight. Photosynthesis fixing carbon dioxide from the atmosphere appears to have evolved very early in the history of life. The 3.8 billion-year-old graphite contains a ratio of carbon isotopes characteristic of the products of photosynthesis. Some scientists argue that there are non-biological ways to make graphite with this isotopic signature. However, 3.5 billion years ago the earliest carbonate sediments have a $\delta^{13}C$ signature that indicates significant organic carbon burial globally, which must have been supported by photosynthesis.

The first photosynthesis was not the kind we are familiar with, which splits water and spits out oxygen as a waste product.

Instead, early photosynthesis was 'anoxygenic'—meaning it didn't produce oxygen. It could have used a range of compounds, in place of water, as a source of electrons with which to fix carbon from carbon dioxide and reduce it to sugars. Potential electron donors include hydrogen (H_2) and hydrogen sulphide (H_2S) in the atmosphere, or ferrous iron (Fe^{2+}) dissolved in the ancient oceans. All of these are easier to extract electrons from than water. Hence they require fewer photons of sunlight and simpler photosynthetic machinery. The phylogenetic tree of life confirms that several forms of anoxygenic photosynthesis evolved very early on, long before oxygenic photosynthesis.

Origin of recycling

The early biosphere fuelled by anoxygenic photosynthesis would have been limited by the supplies of electron donors, all of which are a lot less abundant than water. In fact, shortage of materials would have posed a more general problem for life within the early Earth system. Recall the fluxes of materials coming into the surface Earth system from volcanic and metamorphic processes today (Figure 6). They are many orders of magnitude less than the fluxes due to life at the surface of the Earth today, indicating that today's biosphere is a phenomenal recycling system.

The challenge for early life was to evolve the means of recycling the materials it needed to metabolize—in other words, to establish global biogeochemical cycles. We have a very scant record of how and when this happened, but a few clues suggest it was very early in the history of life. Notably, the carbon isotope record of marine carbonate sediments tells us that the early biosphere was fairly productive, because the fraction of carbon deposited as organic material in sediments, relative to inorganic carbonate, was very similar on the Archean Earth compared to today. Also, the phylogenetic tree of prokaryote life suggests that many recycling metabolisms, such as methane production, evolved early on. The ease or difficulty of evolving recycling has also been explored by seeding computer models with 'artificial life' forms and leaving them to evolve. In these 'virtual worlds' the closing of material recycling loops emerges as a very robust result.

If the early biosphere was fuelled by anoxygenic photosynthesis, plausibly based on hydrogen gas, then a key recycling process would have been the biological regeneration of this gas. Calculations suggest that once such recycling had evolved, the early biosphere might have achieved a global productivity up to 1 per cent of the modern marine biosphere. If early anoxygenic photosynthesis used the supply of reduced iron upwelling in the ocean, then its productivity would have been controlled by ocean circulation and might have reached 10 per cent of the modern marine biosphere.

Origin of oxygenic photosynthesis

The innovation that supercharged the early biosphere was the origin of oxygenic photosynthesis using abundant water as an electron donor. This was not an easy process to evolve. To split water requires more energy—i.e. more high-energy photons of sunlight—than any of the earlier anoxygenic forms of photosynthesis. Evolution's solution was to wire together two existing 'photosystems' in one cell and bolt on the front of them a remarkable piece of biochemical machinery that can rip apart water molecules. The result was the first cyanobacterial cell—the ancestor of all organisms performing oxygenic photosynthesis on the planet today.

Current evidence suggests that oxygenic photosynthesis took up to a billion years to evolve, with the first compelling evidence appearing around 3–2.7 billion years ago. The smoking gun is chemical evidence for oxygen leaking into the environment and reacting with metals that are highly sensitive to the presence of oxygen. For example, molybdenum is mobilized from continental rocks by reacting with oxygen, and appears for the first time in ocean sediments around 2.7 billion years ago. Once oxygenic photosynthesis had evolved, the productivity of the biosphere would no longer have been restricted by the supply of substrates for photosynthesis, as water and carbon dioxide were abundant. Instead, the availability of nutrients, notably nitrogen and phosphorus, would have become the major limiting factors on the productivity of the biosphere—as they still are today.

Once there was a source of oxygen on the planet it is tempting to assume that the concentration of oxygen in the atmosphere would have steadily risen—a bit like filling a bath with the plug in. But oxygen did not rise in the atmosphere immediately or steadily. Instead it remained a trace gas for hundreds of millions of years. We know this because a very peculiar 'mass independent fractionation' (MIF) of sulphur isotopes is preserved in sediments more than 2.45 billion years old. This MIF signature can still be produced by photochemistry of sulphur gases in the atmosphere today, but it cannot be preserved in today's sediments because all the sulphur first goes through a homogenizing reservoir of sulphate in the ocean. Prior to 2.45 billion years ago that sulphate reservoir must have been absent due to a lack of oxygen to produce it. The MIF signature indicates that high-energy ultraviolet (UV) radiation streamed through the lower atmosphere and therefore the ozone layer was absent, requiring that oxygen (from which ozone is made) was at a concentration of less than two parts per million (ppm) in the atmosphere.

Oxygen could remain at such a low concentration for hundreds of millions of years because there was an excess input flux of reduced materials hungry to react with it, including reduced iron injected into the ocean through mid-ocean ridges, and reduced gases such as hydrogen and hydrogen sulphide entering the atmosphere via volcanoes. The rate of their reaction with oxygen increases with oxygen concentration, thus producing a negative feedback system that stabilized the oxygen concentration at the (low) level where the sink of oxygen matched the source. To use the bath metaphor for oxygen in the atmosphere: the plug was out and the plug hole was large, creating a low, stable level for oxygen.

The Great Oxidation

This stability broke down after hundreds of millions of years, when atmospheric oxygen jumped up in concentration in an event known as the 'Great Oxidation' 2.4 billion years ago (Figure 19). The mass independent fractionation of sulphur isotopes stops, indicating that oxygen had risen sufficiently to convert all sulphur to sulphate before it was deposited in marine sediments. The fact that the MIF signature has never returned suggests the permanent formation of an ozone layer. Massive deposits of oxidized iron appeared in the form of the first sedimentary 'red beds'. Rusted (oxidized) iron also appeared in ancient soils for the first time. These indicators reveal that oxygen concentration increased by several orders of magnitude from less than 100,000th to perhaps 1–10 per cent of its present atmospheric level. The indicators of oxygen rise all remain with us to the present day, indicating that the Great Oxidation was never reversed—although some recent work suggests that oxygen may have dipped as low as 0.1 per cent of its present level during the ensuing Proterozoic Eon.

Whilst the origin of oxygenic photosynthesis was ultimately to blame for the Great Oxidation, there must have been other long-term changes in the Earth system that slowly oxidized the surface Earth system. Fundamental amongst these was the loss of hydrogen atoms to space. This is a tiny flux on today's Earth, because water is frozen out of the atmosphere at the 'cold trap' between the troposphere and stratosphere. Hence hardly any hydrogen-containing gases make it to the top of the atmosphere. However, much of the organic carbon created by early oxygenic photosynthesis would have been recycled to the atmosphere as methane. In the resulting methane-rich early atmosphere, much more hydrogen could escape to space and this had the effect of

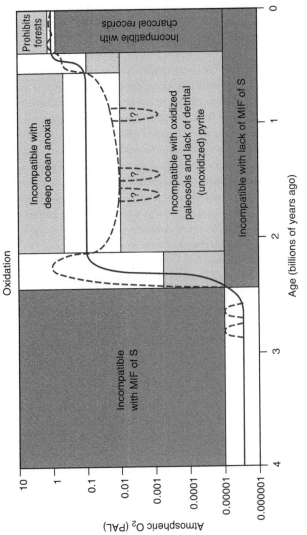

19. Atmospheric oxygen over Earth history.

oxidizing the surface of the Earth. This drove the Earth system towards a tipping point where the balance of inputs to the atmosphere shifted from an excess of reduced material to an excess of oxygen.

The abruptness of the Great Oxidation suggests that at this point a strong positive feedback process kicked in and propelled the rise of oxygen. The formation of the ozone layer was crucial to this transition, because it temporarily suppressed the consumption of oxygen. UV radiation catalyses a series of reactions in which oxygen combines with methane to produce carbon dioxide and water (thus reversing the production of oxygen and methane by the biosphere). Without an ozone layer, this process for removing oxygen was rapid and efficient. But once enough oxygen built up for the ozone layer to start to form, this would have shielded the atmosphere below from UV radiation and temporarily slowed down the removal of oxygen. More oxygen would produce more ozone, letting through less UV radiation and further suppressing oxygen consumption in a positive feedback process. Models suggest that this positive feedback was strong enough to temporarily go into 'runaway', producing a rapid oxygen rise. However, the Earth system would soon have stabilized again at a higher oxygen level with the oxygen sinks again matching the sources.

When oxygen jumped up at the Great Oxidation this caused a decline in atmospheric methane concentration, slowing the further oxidation of the Earth. This decline in methane could help explain why there were a series of glaciations when oxygen levels rose. One of these glaciations, 2.2 billion years ago, reached to low latitudes near the equator and was probably the first snowball Earth event. The Great Oxidation was also followed by a large pulse of organic carbon burial recorded in carbon isotopes. This may have been caused by increased oxygen reacting with sulphide in rocks on the continents producing sulphuric acid that dissolved phosphorus out of rocks and fuelled productivity in the oceans. If so, it reinforced the transition to a higher oxygen world. By 1.85 billion years ago this instability in the carbon cycle and climate had settled down and the Earth entered a long period of stability known rather unflatteringly as 'the boring billion'.

Origin of eukaryotes

The turmoil of the Great Oxidation created a world much more conducive to aerobic (oxygen-utilizing) life forms. There was a lot more energy to go around in the post-oxidation world, because respiration of organic matter with oxygen yields an order of magnitude more energy than breaking food down anaerobically. Amongst the organisms to take advantage of this energy source were the first eukaryotes—complex cells with a nucleus and many other distinct components.

Eukaryotes are profoundly different from the prokaryotes that preceded them, but they are also partly made up of what were once free-living prokaryotes. The mitochondria—the energy factory—in eukaryote cells were once free-living aerobic bacteria, and the plastids in plant and algal cells—where photosynthesis occurs—were once free-living cyanobacteria. These cellular components were acquired in ancient symbiotic mergers with bacteria. The symbiotic merger that gave rise to mitochondria provided an abundant energy source to the ancestral eukaryote cell. Eukaryotes also rearranged how they copy genetic information—copying many chromosomes in parallel—whereas prokaryotes copy their DNA in one long loop. These innovations enabled eukaryotes to express many more genes than prokaryotes, and this ultimately gave them the capacity to create more complex life forms with multiple cell types.

The origin of eukaryotes is shrouded in mystery and controversy, as biologists do not agree on what marks the start of the lineage or what constitutes fossil evidence for eukaryotes. The earliest claims for biomarker evidence of eukaryotes 2.7 billion years ago are now thought to represent contamination with younger material. A couple of cryptic 2.5 billion-year-old 'acritarch' fossils might be the resting stages of early eukaryotes, but the name itself means they are of 'confused origin'. Some 1.9 billion-year-old spiral fossils that are visible to the naked eye might be eukaryotic algae (called *Grypania*) but could also be colonial cyanobacteria. Molecular clocks suggest the last common ancestor of all eukaryotes lived roughly 1.8–1.7 billion years ago.

Eukaryotes only slowly realized their ability to build more complex life forms with differentiated cell types. Most of the fossils from Earth's middle age—the Proterozoic Eon—are the rather cryptic acritarchs. Much rarer eukaryote body fossils include 1.5 billion-year-old *Tappania*, which might be a fungus, and 1.2 billion-year-old *Bangiomorpha pubescens*, which is a multicellular red alga (seaweed) assigned to a modern order.

Researchers are still puzzling over what held back the evolution of complex life during 'the boring billion', but many see environmental constraints playing a key role. For most of the Proterozoic Eon the surface ocean remained dominated by prokaryotes and the deep oceans remained anoxic—i.e. devoid of oxygen. At intermediate depths some of these anoxic waters became 'euxinic', meaning that sulphate in the water was reduced to hydrogen sulphide, which is toxic to many eukaryotes. The peculiar chemistry of the Proterozoic ocean also removed several trace metals such as molybdenum from the ocean. Molybdenum is widely used in nitrogen fixation today, so without it there may have been a shortage of available nitrogen in the ocean.

Neoproterozoic turmoil

Eventually the deadlock was broken in the Neoproterozoic Era (1,000–542 million years ago), which witnessed a spell of climatic turbulence, the oxygenation of the deep oceans, and the rise of the first animals. The first signs of change began around 740 million years ago, when biomarkers of algae become more prevalent in ocean sediments and the diversity of eukaryote fossils starts to increase. This would have made the biological pump of carbon from the surface to the deep ocean more efficient. There were also productive microbial ecosystems on the land at the time, and conceivably eukaryotic fungi, algae, and lichens (a symbiotic merger of the two) could have been part of those early land ecosystems—although we have no fossil evidence either way.

Meanwhile, plate tectonics was breaking up the supercontinent Rodinia and scattering the resulting land masses in an unusual configuration, with much of the land in the tropics. This would have produced very efficient silicate weathering of the continents, potentially enhanced by biology. That in turn would have drawn down atmospheric carbon dioxide and cooled the planet. Somehow the climate got so cold that an extreme glaciation—the Sturtian—was triggered around 715 million years ago. Glaciation reached equatorial latitudes, suggesting this was a snowball Earth event. The glaciation lasted tens of millions of years, consistent with the time it would take to build up enough carbon dioxide to melt the ice.

The climate turmoil did not end there. A second extreme glaciation—the Marinoan—was triggered, ending 635 million years ago. It was followed by a massive deposit of carbonate rock called a 'cap carbonate'—again consistent with the snowball Earth theory (Chapter 1). In the extremely hot and wet aftermath of snowball Earth, weathering would have occurred incredibly rapidly, supplying calcium and magnesium ions to the ocean that would combine with the excess carbon dioxide in the atmosphere and ocean to produce a massive deposit of carbonate sediments.

Perhaps the greatest puzzle about these extreme glaciations is how the ancestors of complex life survived them. Biomarker and molecular clock evidence suggests that simple animals in the form of sponges had already evolved, along with multicellular algae and fungi. Yet complex life did not flourish until after the glaciations. First there are fossils of what are thought to be animal embryos, alongside algae and fungi. Then the first large fossil organisms, the 'Ediacaran biota', appear around 575 million years ago. Whilst their biological affinity is debated, at least some were probably animals. They were followed tens of millions of years later by mobile grazing animals—both on the sediments and as zooplankton in the water column.

What triggered this burst of animal evolution? Relatively large, mobile animals need more oxygen than the sedentary creatures including sponges that came before them. Intriguingly, the first evidence for oxygenation of parts of the deep oceans appears 580 million years ago, shortly before the appearance of Ediacaran fossils at depth in the ocean. However, there had been oxygen in the shallow waters of the ocean for more than a billion years before this. It may be that evolution caused oxygenation rather than vice versa. By increasing the efficiency of carbon removal from the water column and phosphorus removal into sediments, the rise of sponges and algae may have oxygenated the ocean, improving conditions for ongoing animal evolution. The revolution in biological complexity culminated in the 'Cambrian Explosion' of animal diversity 540 to 515 million years ago, in which modern food webs were established in the ocean.

This marked the birth of the modern world. Since then the most fundamental change in the Earth system has been the rise of plants on land (discussed in Chapter 3), beginning around 470 million years ago and culminating in the first global forests by 370 million years ago. This doubled global photosynthesis, increasing flows of materials. Accelerated chemical weathering of the land surface lowered atmospheric carbon dioxide levels and increased atmospheric oxygen levels, fully oxygenating the deep ocean. Since the rise of complex life, there have been several mass extinction events. The end-Permian extinction, 252 million years ago, was the biggest of all, sending the Earth system back towards earlier states, with depletion of the ozone layer and widespread ocean anoxia. However, whilst these rolls of the extinction dice marked profound changes in evolutionary winners and losers, they did not fundamentally alter the operation of the Earth system.

Common features

There have been three revolutionary transformations in the history of the Earth system: the inception of life and biogeochemical cycling; the origin of oxygenic photosynthesis and the Great Oxidation; and the origin of complex life out of the Neoproterozoic environmental turmoil. These revolutions share common features. They were caused by rare evolutionary innovations. They involved step increases in energy capture and material flow through the biosphere, accompanied by increases in the complexity of biological organization and information processing. They relied on the Earth system having some instability, such that new metabolic waste products could cause catastrophic upheavals in climate and biogeochemical cycling. They ended only when blind evolution was able to close the biogeochemical cycles again, recycling the waste materials and establishing a new stable state for the Earth system. Chapter 5 examines whether we humans could be beginning a new revolutionary change in the Earth system.

Chapter 5
Anthropocene

Could the Earth system be on the brink of another revolutionary change, thanks to our activities as a species? We humans are very recent products of evolution, yet already we are transforming the planet at a global scale. The recognition that humans are now a key component of the Earth system was encapsulated in the Bretherton diagram (Figure 5). More recently, the term 'Anthropocene' has been coined to describe a new geological epoch in which human activities are transforming the Earth system at a global scale. There is much debate about whether this really is a new epoch, and if so, when it began. This chapter introduces how human evolution was shaped by changes in the Earth system and how we have gone on to transform the Earth system—tracing the key events on a timeline (Figure 20).

Environmental preconditions

There were several environmental preconditions for human evolution. Perhaps the most obvious is an oxygen-rich atmosphere—our brains are especially energy-hungry and if the partial pressure of oxygen in the air drops by about a third, brain function really starts to suffer. However, we know from the continuous record of fossil charcoal that oxygen has remained above 15 per cent of the atmosphere for the last 370 million years,

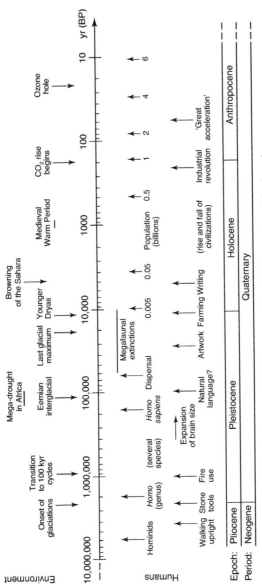

20. Timeline of human evolution set against environmental variability, on a logarithmic scale.

so lack of oxygen has not been holding back our evolution. Instead fires fuelled by abundant oxygen helped create the type of mixed grassland environment in which our ancestors evolved, and fires later became a key tool for early humans.

Although grasslands now cover about a third of the Earth's productive land surface they are a geologically recent arrival. Grasses evolved amidst a trend of declining atmospheric carbon dioxide, and climate cooling and drying, over the past forty million years, and they only became widespread in two phases during the Miocene Epoch around seventeen and six million years ago. These phases of grassland expansion were propelled by a potent positive feedback: grasslands encourage fires which encourage grasslands, because frequent fires prevent forests regenerating. In their second phase of expansion grasslands colonized large parts of Africa, including the Great Rift Valley—the place where our evolutionary lineage diverged from chimpanzees, also around six million years ago. Around four million years ago our hominine ancestors began to walk upright—conceivably as an adaptation to moving through the newly created savannah between clumps of woodland.

Just as our ancestors began to develop stone tool use—first recorded 2.6 million years ago—the Earth descended into a series of ice age cycles of increasing severity and decreasing frequency. This change in climate dynamics marks the onset of the Quaternary Period. It provoked widespread speciation in mammals, including our hominine lineage. The resulting global environmental instability may have played a role in our evolution as unusually intelligent, highly social primates. The general idea is that when the environment is changing—but not too frequently or unpredictably—it pays to be smart and to cooperate in social groups to help adapt to the changing conditions. In contrast, if the environment is stable there is no need to be clever, and if the environment is really volatile the best strategy is to just breed your way out of trouble.

Fire use

The intentional use of fire set our ancestors apart from all other species, because it was the first innovation that extended energy use beyond the human body. Controlled use of fire may have started 1.5 million years ago, and was certainly occurring by 800,000 years ago in Africa and by 400,000 years ago in Europe. The use of fire for cooking gave *Homo erectus* more energy in their diet from the cooking of meat and a more diverse diet (by detoxifying foods). The shift to hunting energy-rich meat in turn triggered the formation of social groups that settled around campsites and divided labour, causing an escalation in human social evolution.

Between about 400,000 and 250,000 years ago, stone tool technologies became more elaborate and brain size increased rapidly. Anatomically modern humans (*Homo sapiens*) first appeared in East Africa around 200,000 years ago. Sometime after that, our ancestors experienced a bottleneck in population of 10,000 or fewer breeding pairs. Descendants of this founding group emerged out of Africa and began to spread around the world roughly 65,000 years ago. Their migration was facilitated by one of a series of periodic wet phases of the Sahara after a mega-drought in Africa from 135,000 to 90,000 years ago. As modern humans arrived in new continents, they triggered the extinction of other large mammals or 'megafauna'. This began in Australia before 44,000 years ago, in Europe over 30,000 years ago, in North America 11,500 years ago, and in South America 10,000 years ago. Extinction was less severe in Africa, perhaps because existing species were already habituated to and wary of human hunters.

Fire was the first 'tool' that enabled early humans to start changing their environment on a large scale. Human use of fire in hunting shifted ecosystems towards grasslands. This helps explain why herbivores that browse on trees (rather than eating grasses) suffered most in the megafauna extinctions. Our ancestors may also have hunted some large herbivores to extinction, thus leaving carnivores and scavengers to suffer from a lack of food. Human use of fire in Australia helped maintain desert scrubland over large areas of the continent. This in turn may have inhibited the return of the monsoon into the continental interior when the Earth system entered the present Holocene interglacial epoch. If so, it may represent the first large-scale impact of humans on the climate system.

Farming

As the Earth system exited the last ice age there was a major fluctuation in the climate of the Northern Hemisphere. An abrupt warming around 14,700 years ago was followed by a marked cooling 12,700 years ago and a further abrupt warming 11,500 years ago. During the cool period known as the 'Younger Dryas', people in the Eastern Mediterranean region who had been collecting abundant wild cereals for food began domesticating the first cereal crops, perhaps in response to the regional drying effects of climate change. As the Earth system settled into the stable Holocene interglacial state, around 10,500 years ago, the Sahara re-entered one of its wet and green phases, turning the region encompassing the Nile, Euphrates, and Tigris rivers into the fabled Fertile Crescent. Farming began there with the domestication of wheat, barley, peas, sheep, goats, cows, and pigs. Farming also arose independently elsewhere in the world, around 8,500 years ago in South China, 7,800 years ago in North China, 4,800 years ago in Mexico, and 4,500 years ago in Peru and eastern North America.

The relatively abrupt and independent occurrence of farming all over the world suggests it may have been held back by environmental conditions before the Holocene. Low ice age levels of carbon dioxide and the volatile glacial climate would certainly not have helped establish agriculture. Once established, farming increased the energy input to human societies. This 'Neolithic revolution' caused an increase in human fertility (soon followed by an increase in mortality), which increased the population from six million to over thirty million between 6,000 and 4,000 years ago and perhaps as high as 100 million by 2,000 years ago. However, one of the downsides of farming was that sedentary, high-density agricultural civilizations were more sensitive to climate change than mobile foraging societies—with abrupt shifts in tropical climate during the Holocene linked to the collapse of several ancient societies.

The increased population and energy flows due to farming were linked to an increase in material inputs to, and waste products from, societies. The resulting environmental effects began early in the Holocene, but their scale is much debated. Irrigation began around 8,000 years ago in Egypt and Mesopotamia, with the diversion of flood waters from the Nile and the Tigris/Euphrates. This led to some salination and siltation of the land, reducing crop yields and encouraging a shift in agricultural crop from wheat to more salt-tolerant barley. Fertilization of cropland through the use of minerals or manure was practised by the Egyptians, Babylonians, and Romans and presumably had knock-on effects in neighbouring freshwaters. Soil erosion is remarked on by Plato, who likened the land to 'the skeleton of a sick man, all the fat and soft earth having wasted away, and only the bare framework of the land being left'. But did the onset of farming impact the Earth system at a global scale?

The early Anthropocene hypothesis

Bill Ruddiman argues that the Anthropocene began thousands of years ago, as a consequence of the Neolithic revolution. The associated population expansion certainly drove forest clearance to create agricultural land and supply biomass energy and wood as a material. Forest clearance in turn reduced the carbon storage capacity of the land, transferring CO_2 to the atmosphere. Ruddiman argues this effect was so great that from 8,000 years ago onwards this CO_2 source was enough to outweigh what should have been a natural decline in atmospheric CO_2. Furthermore, from 5,000 years ago onwards, irrigation of rice paddies produced a source of methane that Ruddiman argues outweighed an expected decline in atmospheric CH_4.

Other researchers have used Earth system models to show that natural changes in the climate and carbon cycle can explain most of the changes in atmospheric CO_2 and CH_4 during the Holocene. For example, variations in the Earth's orbit meant that 6,000 years ago the Northern Hemisphere was warmer than today and therefore supported more vegetation, both in the boreal regions and across much of North Africa—creating a 'green Sahara'. This helps explain somewhat lower atmospheric CO_2 levels in the early Holocene. As the orbital forcing steadily declined there was a relatively abrupt drying and expansion of the Sahara desert, around 5,000 years ago. Models predict this was due to a shift between alternative steady states of the vegetation-climate system in North Africa. This 'browning of the Sahara' together with a retreat of boreal forests from the highest Northern latitudes added CO_2 to the atmosphere.

Over the past two millennia, our records of past climate change improve, with multiple proxies for climate variability including tree ring and ice core records and temperatures from boreholes. These records reveal slow fluctuations between somewhat warmer and cooler intervals on Northern Hemisphere land surfaces, including the Medieval Warm Period (c.950–1250 AD) and the Little Ice Age (c.1550–1850 AD). Intervals of cooler climate correlate with poor agricultural production, war, and population decline, but any causal links are argued over. Ice core records reveal some variations in atmospheric composition including a 10 ppm decline in CO_2 500 years ago, which was also a time when human biomass burning declined. Bill Ruddiman argues this was due to plague-induced human population decline which allowed large areas to reforest and take up carbon. However, his 'early Anthropocene' hypothesis continues to be controversial, partly because pre-industrial societies were limited in the energy supplies with which they could transform their environment.

Fossil fuels

Most researchers link the start of the Anthropocene with the Industrial Revolution, because the accessing of fossil fuel energy greatly increased the impact of humanity on the Earth system. The Industrial Revolution marks the transition from societies fuelled largely by recent solar energy (via biomass, water, and wind) to ones fuelled by concentrated 'ancient sunlight'. Although coal had been used in small amounts for millennia, for example for iron making in ancient China, fossil fuel use only took off with the invention and refinement of the steam engine. Thomas Newcomen's demonstration of a working steam engine in 1712, followed by James Watt's improvements to it in 1769, gave a great boost to coal extraction, by draining mines of water. The steam engine was also used to convert fossil fuel energy into mechanical power in manufacturing and transport. This created a potent positive feedback loop that propelled the Industrial Revolution.

The exploitation of concentrated fossil fuel energy (Figure 21) triggered a massive expansion of population, food production, material consumption, and associated waste products. Human population doubled between 1825 and 1927 from one to two billion, doubled again by 1975 to four billion, and is on course to double again by 2030 to eight billion. With the Industrial Revolution, food and biomass have ceased to be the main source of energy for human societies. Instead the energy contained in annual food production, which supports today's population, is at fifty exajoules (1 EJ = 10^{18} joules), only about a tenth of the total energy input to human societies of 500 EJ/yr. This in turn is equivalent to about a tenth of the energy captured globally by photosynthesis.

The corresponding increase in global material flows—the largest of which is carbon dioxide (Figure 21)—is disrupting the Earth system. Material waste products have been dumped on land, in the atmosphere, and in the ocean. For some elemental cycles our collective activities now exceed the activities of the rest of the biosphere combined. Much of this escalation of human impact on the Earth system has occurred since the end of World War II—in a transition dubbed the 'Great Acceleration'. The following sections detail some of these changes in material flows and their consequences.

Changes in land-use and nutrient cycles

The increase in food production that now supports over seven billion people has been propelled by increasing inputs of land, nutrients, herbicides, and fossil fuel energy. The second and third billion people were added largely by increasing the area of land under cultivation, helped by replacing horses with tractors, increasing irrigation, and adding herbicides. The fourth and fifth billion were added by a dramatic increase in fertilizer nutrient inputs to existing land, complemented by the introduction of dwarf varieties of wheat and rice, which could thrive on the high-nutrient inputs. The sixth and seventh billion were added largely through increases in crop yield in developing nations based on the spread of earlier innovations.

The expansion and intensification of agriculture has changed the visible face of the Earth. Arable cropland expanded from around 0.5 billion hectares in 1860 to nearly 1.4 billion hectares in 1960. Since then there has been little change in arable land area, but increasing meat consumption in the average human diet has driven an expansion of rangelands for grazing to over three billion

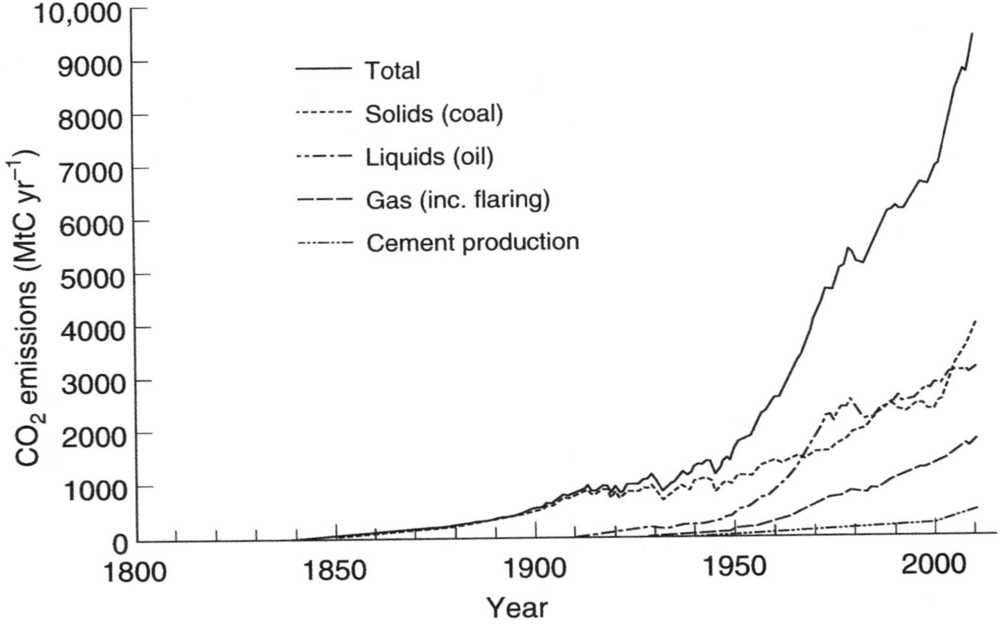
10,000
9000
8000
7000
6000
5000
4000
3000
2000
1000
0
CO_2 emissions (MtC yr^{-1})
Total
Solids (coal)
Liquids (oil)
Gas (inc. flaring)
Cement production
1800
1850
1900
1950
2000
Year

21. **Escalating human fossil fuel CO_2 emissions.**

hectares today—and this has been one of several drivers of rapid tropical deforestation.

The intensification of agriculture has also transformed global nutrient cycling. Fossil fuel energy is used to split the triple bond of N_2 and make nitrogenous fertilizers, and to mine and refine phosphorus fertilizer. This has roughly doubled the input of available nitrogen to the biosphere and tripled the input of phosphorus. Whilst this 'green revolution' has helped protect terrestrial ecosystems from the plough, it has had other negative consequences. Much of the nitrogen and phosphorus we add ends up in freshwaters where it fuels biological productivity (eutrophication), sometimes to the extent that ancient cyanobacteria choke out more recent life forms, with waters turning anoxic and killing fish and other animals. Some of our added nitrogen and phosphorus reaches coastal seas and ultimately the open ocean, driving those waters towards anoxia.

A fraction of the nitrogen that humans have synthesized and added to agricultural soils has been converted to the long-lived greenhouse gas nitrous oxide, by the ancient microbial processes of nitrification and denitrification. This has increased the atmospheric concentration of nitrous oxide from 272 to 310 ppb. Expanding agriculture has also increased methane emissions to the atmosphere, especially from ruminant livestock and paddy fields. Together with leaks of natural gas during its extraction, transport, and use, and emissions from landfills, fires, and waste treatment works, human activities have caused the methane concentration in the atmosphere to more than double, from around 800 ppb to around 1,800 ppb today.

Carbon cycle change

Prior to the Industrial Revolution, the massive exchange fluxes of CO_2 between the atmosphere and ocean, and between the atmosphere and land, were approximately in balance. Since then the emissions of CO_2 from burning fossil fuels (Figure 21) and from land-use change have increased the concentration of CO_2 in the atmosphere from 280 ppm to around 400 ppm today. By the time Dave Keeling started measuring atmospheric CO_2 in 1958 it had already risen to 315 ppm. The 'Keeling curve' (Figure 22) has revealed an accelerating rise of CO_2 since then. Yet CO_2 is not rising as fast as it is being added to the atmosphere. The reason is that about half of the CO_2 added annually is being taken up by ocean and land 'carbon sinks'.

The ocean carbon sink exists because gaseous CO_2 dissolves in and then reacts with seawater. Adding excess CO_2 to the atmosphere forces some of it to dissolve in the surface ocean. The exchange of gas across the sea surface is relatively quick, but dissolved CO_2 then reacts more slowly with seawater. Adding a reactant to one side of a chemical reaction always drives the reaction to the other side, until a new equilibrium is achieved—thus CO_2 is transformed into dissolved inorganic carbon. In fact there are a series of reactions and at equilibrium the stable place for most of the carbon to be is dissolved in the ocean, not as CO_2 in the atmosphere. However, the well-mixed surface layer of the ocean has a relatively small volume, so the rate of ocean carbon uptake is limited by the relatively slow exchange of surface waters with the bulk of the deep ocean.

The land carbon sink exists because those ecosystems—especially forests—that are not being cleared for agriculture are accumulating carbon in live vegetation and soils. One key reason is that increasing atmospheric CO_2 fertilizes photosynthesis—making the uptake of carbon by plants more efficient. This is because there is a competition between CO_2 and O_2 molecules for the active sites of the carbon-fixing enzyme RuBisCO, so increasing the CO_2/O_2 ratio increases the amount of CO_2 that gets fixed. In addition to this 'CO_2 fertilization effect' some ecosystems are being fertilized by human inputs of nutrients, often carried in gaseous form by the atmosphere. Also, where agricultural land is abandoned natural

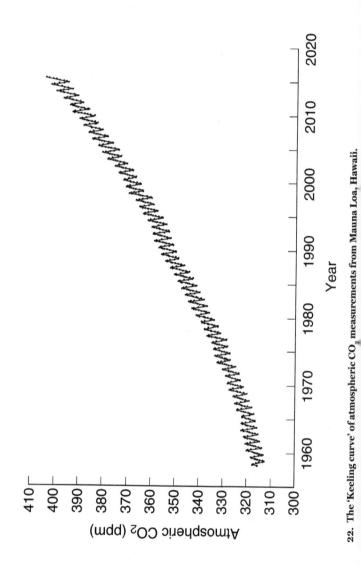

22. The 'Keeling curve' of atmospheric CO_2 measurements from Mauna Loa, Hawaii.

ecosystems tend to regrow and accumulate carbon from the atmosphere.

The fluctuations in the Keeling curve hide a wealth of additional information. They are particularly revealing about the land carbon sink, which is much more variable from year to year than the ocean carbon sink. Overlaid on the overall rise in atmospheric CO_2 is an annual cycle representing the seasonal 'breathing' of the terrestrial biosphere—CO_2 declines in boreal spring and summer as plants in the Northern Hemisphere take up carbon, and rises again in autumn and winter as the same ecosystems exhale CO_2. However, the size and shape of this oscillation varies from year to year. After the eruption of Mount Pinatubo in 1991, CO_2 rose more slowly because the resulting cooling strengthened the land carbon sink. After the strong El Niño of 1998, CO_2 rose faster because warming that year and associated fires eliminated the land carbon sink.

There are positive feedbacks between climate change and the carbon cycle (Chapter 3)—as temperature goes up, the land becomes a less effective carbon sink. So too does the ocean, because warming makes CO_2 less soluble. Furthermore, the uptake of CO_2 is acidifying the ocean and this makes it a less effective store of carbon (by shifting the equilibrium of the reactions with seawater towards gaseous CO_2). Overall, however, negative feedback in the carbon cycle wins out and is slowing the rise of atmospheric CO_2. Were it not for the land and ocean carbon sinks, CO_2 would already be above 500 ppm of the atmosphere, and there would have been greater climate change.

Climate change

The theory of the greenhouse effect is Victorian physics. As early as 1896, Svante Arrhenius calculated that a doubling of atmospheric CO_2 from its pre-industrial concentration would warm the world by around 5°C. This laborious calculation by hand, which took him two years, remains within the range of estimates of 'climate sensitivity' from the latest Earth system models. The current best estimate is around 3°C.

By the end of the 19th century, ship-borne temperature measurements were also being regularly made. These, together with thermometer readings from land-based weather stations, have enabled climate scientists to piece together what is called the 'instrumental temperature record' (Figure 23). It shows global warming of around 0.85°C from 1880 to 2012, around 0.5°C of which has occurred since 1980. The rise in temperature is global in extent (whereas the Medieval Warm Period and Little Ice Age were only regional phenomena). The global temperature has not risen at a steady rate—there are some periods of stable temperature (e.g. the 1940s and 1950s) and some spells of more rapid rise (e.g. the 1980s and 1990s). This is to be expected because even in the absence of human activities there is natural variability in the climate, producing warmer and colder spells—which when superimposed on a rising trend gives periods of no warming and periods of rapid warming.

One factor that can cool the climate is the injection of tiny reflective sulphate aerosol particles into the atmosphere, which scatter sunlight (sending some of it back to space). Sulphate aerosols can come from volcanic eruptions (such as Mt Pinatubo in 1991), or from fossil fuel burning, especially the combustion of sulphurous (brown) coals. However, when it enters into solution, sulphate forms sulphuric acid and hence acid rain. In order to curb acid rain, successful efforts have been made to scrub sulphur dioxide out of power station flue gases. This in turn reduced its cooling effect on the climate, unmasking the increasing greenhouse effect, and contributing to global warming.

Human planet

Our species evolved amidst an unusually unstable climate, spreading worldwide and domesticating the first crops and

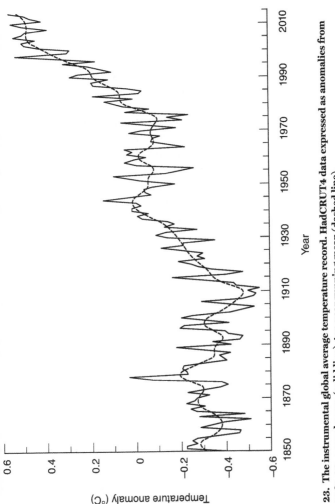

23. The instrumental global average temperature record. HadCRUT4 data expressed as anomalies from 1961–90: annual mean (solid line), ten-year running mean (dashed line).

livestock. In the relative stability of the present Holocene interglacial, farming as a way of life started to take over from hunter-gathering, and the first city states emerged. Humans began to alter the land surface and thus the carbon cycle, the climate, and other biogeochemical cycles. Locally, civilizations started to reach the limit of natural resources, and their fate was occasionally sealed by natural shifts in the climate. At some point—much debated—humans began to alter the entire Earth system. With the Industrial Revolution, human reshaping of the Earth system accelerated. In the technological optimism of the 1950s a further 'great acceleration' occurred. Yet out of that optimism emerged the space race and the dawning realization of the finite beauty of our planetary home. Now humans are dominant players in the biogeochemical cycles of phosphorus, nitrogen, and carbon. We are altering the climate, massively accelerating erosion on land and sedimentation in the ocean, acidifying and deoxygenating the ocean, and extinguishing other species at an unprecedented rate. Chapter 6 considers where this trajectory is taking us.

Chapter 6
Projection

Where is the Earth system heading in the Anthropocene? To even begin to answer this question requires a model of how the Earth system works, and the answer depends on our collective activities as a species, and how the Earth system responds to those. The model's role is to forecast the consequences of different assumptions about future human activities. This chapter introduces 'Earth system models' and some of the crucial assumptions that go into using them to forecast the future. It outlines their projections, going from shorter to longer timescales, and from the specific challenge of projecting climate change to the broader challenge of exploring other global changes.

Earth system models

An Earth system model is a representation of the surface Earth system in a computer program. Like all systems, the model's boundaries have to be carefully defined. In the current generation of models, human activities are treated as an input to the model, as if from outside—even though we clearly live within the Earth system. The model captures the non-human parts of the Earth system, including the atmosphere, ocean, land surface, marine and terrestrial biosphere, and the interactions between them, including the (short-term) carbon cycle.

The most complex Earth system models started life as weather forecasting models. Over the last few decades they have been transformed from models of the atmosphere to models of the Earth system, by progressively adding components—effectively expanding the system being considered. Each time a component is added, a new set of feedbacks is introduced and the results are not always stable. Notably the first models that coupled the slow dynamics (and large heat capacity) of the ocean to the fast dynamics of the atmosphere had a habit of drifting away from anything resembling the current climate state, and had to be artificially dragged back towards observations by 'flux correction'—a problem that was only solved in the 1990s.

What currently marks out an 'Earth system' model is the ability to translate human activities, such as the emissions of greenhouse gases and aerosols, into their climatic effects. The first models to achieve this were published around the year 2000 and included an interactive global carbon cycle that could calculate the effect of CO_2 emissions on the concentration of CO_2 in the atmosphere, including the exchange of CO_2 with the ocean and land ecosystems. Subsequently some models have included the effects of sulphur dioxide emissions on sulphate aerosol formation and hence cloud properties and climate. The latest models also include the effects of changing human land-use on the climate.

Testing the models

A key test of Earth system models is their ability to reproduce already observed climate change. This test typically involves forcing a model with known natural and human climate drivers over the last 150 years, for which we have an observational record of the climate state. This has revealed that both natural and anthropogenic forcing factors are required to reproduce the instrumental temperature record (Figure 23). Natural factors alone cannot produce the marked warming over the last half century—in fact they lead to a predicted slight cooling.

Anthropogenic factors are responsible for the overall warming trend but they cannot explain some of the fluctuations in the temperature record, which are due to natural factors. One of these fluctuations is the slowdown in atmospheric warming over the last fifteen years which is due to an increase in the efficiency of heat uptake by the ocean. In fact most of the heat trapped by the enhanced greenhouse effect goes into the oceans, which have a far greater capacity to store it than the atmosphere. So, we should not be surprised that fluctuations in ocean heat storage affect the temperature of the atmosphere.

If they are given CO_2 emissions as an input, Earth system models do a reasonable job of predicting the historical rise in CO_2 concentrations. Another important test of a model is its ability to capture past climate changes before the observational record. However, state-of-the-art models are so expensive to run that such tests have been rather limited. One area where they really struggle is in capturing past abrupt climate changes.

A spectrum of models

The desire to understand past as well as future global changes has led to the generation of a spectrum of Earth system models of varying complexity. The most sophisticated models—just described—are focused on the challenge of predicting climate change on a relatively short (century) timescale. They resolve the Earth in three dimensions at as high a spatial resolution as the fastest supercomputers will allow, but exclude longer-term processes such as interaction with the Earth's crust. 'Intermediate complexity' models are designed to simulate longer timescales, from a thousand to a million years, and can include the weathering of rocks on the continents and the deposition of sediments in the ocean. These models resolve the ocean and atmosphere at a lower spatial resolution, often simplifying their physics and sometimes reducing their dimensionality—for example representing the ocean as a series of two-dimensional slices with depth and latitude. Then there are simple models with very limited spatial resolution that capture aggregate variables like the global average surface temperature, but can include more parts of the Earth system. These can serve several purposes: to simulate geologic timescales, to be run millions of times to explore the sensitivity of results to uncertain assumptions, or to form part of an 'integrated assessment model'.

Integrated assessment models are primarily targeted at exploring policy options for tackling climate change. They focus on having a simple representation of the economy, including how it currently generates fossil fuel emissions, coupled to a simple model of the climate system, which feeds back impacts on the economy. Typically a decision maker is imagined who has the power to change policy, for example by setting a tax on CO_2 emissions. The model gives an answer to what is the optimal policy now and in the future, given the modeller's assumptions about the costs and benefits of different courses of action. To solve such a computational problem typically assumes that the policymaker is a rational agent with perfect knowledge of the future consequences of their actions (although some models do include uncertainty about the future). There is much debate about this approach and the assumptions that go into it. Yet despite their simplicity, integrated assessment models represent the first attempt to model humans as an interactive part of the Earth system.

Projection rather than prediction

Projecting climate change (or any other long-term global change) is fundamentally different from predicting the weather. Predicting the weather is an 'initial conditions problem'—meaning that the future state of the weather depends fundamentally on its present (and past) state, which therefore has to be put into a model as accurately as possible. Even then the weather can be so sensitive to initial conditions that small differences rapidly lead to very different outcomes. This is a classic example of the phenomenon of 'deterministic chaos' first described by Ed Lorenz with a simple three-equation model of the atmosphere in 1963.

The climate on the other hand is defined as the long-term average of the weather (typically over a thirty-year period) and it is not so sensitive to initial conditions, because only inherently 'slow' parts of the Earth system such as the ocean can carry a memory of the initial conditions for that long. Instead, predicting the climate is more of a 'boundary conditions problem'—meaning that it depends on factors such as the Earth's orbit and the levels of different greenhouse gases and aerosols in the atmosphere. At intermediate, seasonal to decadal timescales the 'memory' of initial conditions carried especially by the ocean is important to accurate prediction. Hence a major effort has recently gone into initializing decadal climate predictions with observations of the present ocean state.

The further we try to forecast climate change into the future, the more the forecast depends on the trajectory of key forcing factors such as CO_2 emissions. The fundamental issue with these human-influenced factors is that they have yet to be determined. No one pretends that they can 'predict' how future societies are going to develop on a century timescale. Instead all we can do is come up with a range of storylines about how societies and their emissions may evolve, and use those as inputs—'forcing scenarios'—to Earth system models. For this reason it is more accurate to describe the output of the models as climate change 'projections'—which depend on their stated assumptions—rather than 'predictions'.

Storylines

The default storyline about the future is an extrapolation of 'business as usual', implying ongoing growth in fossil fuel consumption. Currently CO_2 emissions are around ten billion tonnes of carbon per year (Figure 21). They grew at around 2 per cent per year averaged over the last thirty years, but closer to 3 per cent per year over the last decade—despite the global recession. If growth rates of 2–3 per cent per year continue this will double emissions within another 25–35 years—i.e. by mid-century. However, such exponential growth produces implausibly high numbers if we extrapolate it to the end of the century. Ultimately it must be constrained by the fact that fossil fuels are a finite resource. Nevertheless, business-as-usual scenarios typically project a roughly three-fold increase in CO_2 emissions by the end of this century. Earth system models predict this would increase the atmospheric concentration of CO_2 above 1,000 ppm—from a pre-industrial level of 280 ppm.

A very different storyline is that we will act collectively and decisively to stop the rising trend of CO_2 emissions and then to reduce them—a 'strong mitigation' scenario. Such scenarios are constructed to illustrate what could be gained by taking decisive action to tackle climate change, and often have a target behind them, such as limiting global warming to less than 2°C. Alas they have tended to become rapidly out-of-date as actual emissions have risen exponentially. Typically, strong mitigation scenarios show global emissions falling to less than half their present level by mid-century and then continuing to decline towards zero.

An important take-home message from Earth system models is that CO_2 emissions will have ultimately to be reduced to zero to stop the atmospheric CO_2 concentration from rising. In the short term, atmospheric CO_2 concentration can be stabilized by reducing emissions to match the flux of CO_2 to the deep ocean, which is around 10 per cent of present emissions. This could stabilize the atmospheric CO_2 concentration at around 560 ppm (twice the pre-industrial level) or above. However, the most optimistic scenarios seek to stabilize atmospheric CO_2 concentration at around 450 ppm on the century timescale. Given recent growth in CO_2 emissions, these scenarios to limit global warming to 2°C now require that by the end of this century societies are deliberately removing CO_2 from the atmosphere.

In between these two extremes there are a range of socio-economic scenarios with different assumptions about growth in energy demand or effort to mitigate CO_2 emissions. These reflect storylines such as continued globalization or descent into a more politically fragmented world. Typically, integrated assessment models are used to generate these scenarios.

The ultimate constraint on CO_2 emissions will be that there is only so much fossil fuel in the ground. However, the amount is large—at least 5,000 billion tonnes of carbon—and the estimates have got larger of late, as extraction methods improve. If the price of fossil fuel rises, the size of the reserves that it becomes economical to extract increases. But so too does the incentive to switch to other cheaper energy sources. Beyond the century timescale, some scenarios are constrained by an estimate of total fossil fuel reserves, providing an instructive indication of how much we could alter the climate in the long term. These are best viewed as 'thought experiments'.

Global warming

Projections of global temperature change depend fairly linearly on the cumulative emissions of CO_2 up to a given time, i.e. how much fossil fuel we burn without capturing and storing the CO_2 given off (Figure 24). Roughly speaking, every 500 billion tonnes of carbon emitted will give 1°C of global warming. Thus, we have burned around 400 billion tonnes of fossil fuel carbon already and have experienced 0.8°C of warming. If we want to stay under 2°C of warming we need to limit our emissions to a trillion (1,000 billion) tonnes of carbon. Whereas if we burn all 5,000 billion tonnes of known fossil fuels we can eventually expect around 10°C of warming. Whether this thought experiment can ever be realized

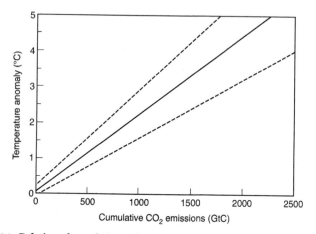

24. Relation of cumulative carbon emissions to global temperature change from a range of recent Earth system models.

is highly questionable because < 10°C warming could be so damaging as to prevent us from burning all the fossil fuel.

On the short timescale of the next few decades, temperature projections are not very dependent on the emissions pathway, because the climate system is still responding to the energy imbalance caused by the accumulation of past greenhouse gas emissions. Also, natural variability in heat uptake and storage by the ocean can affect surface temperatures considerably on decadal timescales.

On the long timescale of a millennium, temperature change still depends on the total cumulative emissions of carbon. However, by then the Earth system will have apportioned the CO_2 we have added between the atmosphere, ocean, and land surface. The fraction remaining in the atmosphere—known as the 'airborne fraction'—will depend on the total amount of carbon we emit. At a minimum it will be about 20 per cent. But simple and intermediate complexity models tell us that the fraction increases exponentially with the amount of carbon added. As temperature depends on the natural logarithm of atmospheric carbon, these two effects combine to give a linear relationship between carbon emitted and global warming.

The relationship between carbon in the atmosphere and global temperature change is captured in a concept called 'climate sensitivity'. This is defined as the global warming caused by doubling the CO_2 content of the atmosphere, once the heat content of the ocean has adjusted and various 'fast' feedbacks have operated. Our best estimate is that it is near 3°C, but it could be in the range 1.5–5°C. It is uncertain because our models differ over the strength of different feedbacks, and over the long-term heat uptake by the deep ocean, and observations cannot completely constrain these properties. To the 'climate sensitivity' we can add the sensitivity of atmospheric CO_2 concentration to a given CO_2 emission, which depends on feedbacks between the climate and the carbon cycle. On longer timescales there are further 'slow' feedbacks, for example involving the melt of ice sheets, which add to warming. The resulting 'Earth system sensitivity' to CO_2 could be as much as twice the climate sensitivity.

Spatial patterns

Climate change is not spatially uniform. The Arctic is already warming twice as fast as the global average, and further 'polar amplification' of warming is forecast by models. Warming is also more rapid on land than over the ocean, because the land has a much lower capacity to soak up heat—whereas the ocean can store it. In practical terms this means the warming in continental interiors can be twice the global average and Arctic land masses may warm three times the global average.

The spatial pattern of changes in the water cycle is harder to forecast than changes in temperature. Warming will increase evaporation from the ocean and a warmer atmosphere can hold more moisture—this is 19th-century physics known as the Clausius–Clapeyron equation. Sure enough, models predict that a warmer atmosphere will be a wetter one, but how much wetter is uncertain. The water cycle can be pictured as spinning faster in a warmer world with models generally predicting wet areas getting wetter, whereas some dry land areas are forecast to get drier. The great circulation cells of the atmosphere—the Hadley cells—which rise at the equator and descend in the tropics are forecast to expand, spreading regions of dry descending air polewards, and intensifying some dry regions such as the Mediterranean. Such spatial patterns of climate change will be crucial to determining the response of key parts of the Earth system—as well as the impacts on us humans.

Tipping points

Whilst much of the behaviour of the Earth system can be described as 'linear' and predictable with our current models, there is a class of 'non-linear' change that is much harder to predict and potentially much more dangerous. It involves 'tipping points'—where a small perturbation triggers a large response from a part of the Earth system—leading to abrupt and often irreversible changes. Tipping points can occur when there is strong positive feedback within a system, which creates alternative stable states for a range of boundary conditions. When changes in the boundary conditions cause the current state of a system to lose its stability, a tipping point occurs, triggering a transition into the alternative stable state. Thankfully it is very difficult to pass a tipping point at the planetary scale—rare examples from Earth history are the switches into (and out of) 'snowball Earth' (Chapters 1 and 4).

However, several subsystems of the Earth system are thought to exhibit alternative stable states and tipping points. An archetypal example is the Atlantic Ocean's overturning circulation. I have dubbed those parts of the Earth system that can exhibit tipping points 'tipping elements'. Amongst them are several candidates that could be tipped by human-induced global change (Figure 25). They can be divided into those involving abrupt shifts in modes of circulation of the ocean or atmosphere (or the two of them coupled together), those involving abrupt shifts in the biosphere, and those involving abrupt loss of parts of the cryosphere.

Ocean and atmosphere

The circulations of the atmosphere and ocean are coupled together and have undergone abrupt changes in the past. The Atlantic Ocean's overturning circulation comprises a northward flow of surface waters from the South Atlantic, across the equator, and up to the Northernmost Atlantic where the water becomes dense enough to sink to depth, supporting a deep return flow southward. This circulation is self-sustaining thanks to a positive feedback whereby it imports salt from the Southern Ocean, thus making Atlantic waters denser and more prone to sinking. However, if the 'boundary conditions' are changed, by adding extra freshwater to the North Atlantic, a tipping point can be reached where the formation of deep waters ceases. Then the overturning circulation stops, and it enters a stable 'off' state. To recover the 'on' state of overturning circulation then requires a much larger reduction in freshwater input.

Switches between alternative stable states of the Atlantic overturning circulation are implicated in both past rapid North Atlantic warming events (abrupt strengthening of the circulation) and rapid cooling events (abrupt collapse of the circulation). Already more freshwater is entering the North Atlantic region due to increased rainfall, and models project a weakening of the overturning circulation. In some business-as-usual model scenarios, the overturning circulation eventually collapses, with knock-on effects around the planet.

In the past, strengthening or weakening of the Atlantic's overturning circulation caused the intertropical convergence zone

25. Map of potential tipping elements in the Earth's climate system.

of rainfall to move northwards or southwards, sometimes triggering abrupt shifts in the monsoons in West Africa and India. A monsoon can be thought of as an overturning circulation of the atmosphere with moist air drawn in from the ocean over the continent, where it rises and cools, causing water to condense and precipitate, thus releasing latent heat that drives the upward convection of air—a strong positive feedback that supports the monsoon circulation. Monsoons are driven seasonally by the land warming up faster than the ocean, and their seasonal switching on and off supports the idea that monsoons exhibit tipping points. Some future projections show abrupt shifts in the monsoons, for example in West Africa, where the warming of the waters offshore could tie rainfall to the coast, starving the Sahel of its seasonal supply of water.

Land biosphere

Some parts of the land surface are strongly coupled to the atmosphere through positive feedbacks. For example, the 'green Sahara' state that was present 6,000 years ago supported an atmospheric circulation that brought moisture into what is now a desert. Today the Amazon rainforest recycles water to the atmosphere, thus helping maintain the rainfall that supports the forest. It also suppresses fires. However, if the climate dries regionally—as has been seen in recent Amazon drought years (2005, 2010)—this can lead to dieback of trees and a shift to a more devastating fire regime. If grasses begin to encroach into the forest these encourage fires which destroy tree saplings and support an alternative grassland or savannah state (a positive feedback). Grasslands are already thought to be an alternative stable vegetation state for large parts of the Amazon basin under present rainfall. In the future, if the region dries out, widespread dieback of the Amazon rainforest has been projected.

Elsewhere, several regions of boreal and temperate forests are already experiencing widespread dieback thanks to bark beetles that are thriving in a warmer climate. In some future projections, large areas of boreal forest are lost due to bark beetle attacks, increased fires, and summers getting too hot for the trees. Dieback of the Amazon or boreal forests would in turn feedback CO_2 to the atmosphere, but their potential contribution is modest compared to projected human CO_2 emissions.

Cryosphere

The amplified warming of the Arctic is partly due to the ice-albedo positive feedback that we met in Chapter 1—as sea-ice is lost, this exposes a dark ocean surface that absorbs more sunlight. The loss of Arctic sea-ice has been accelerating and complete summer ice loss is forecast in the next few decades. If we continue with business as usual, then models project year-round loss of the Arctic sea-ice in the next century. In some models this year-round ice loss happens very abruptly when winter temperatures fail to reach the freezing point across a large part of the Arctic Ocean.

Warming of the Arctic land surfaces is already thawing permafrost—frozen soils—and releasing the reservoirs of methane and carbon dioxide they contain. Under business as usual, most of the permafrost is projected to be lost by the end of this century, amplifying global warming by about 10 per cent. On a longer timescale, ocean warming will destabilize frozen reservoirs of methane (known as hydrates or clathrates) underneath ocean sediments. The resulting out-gassing of carbon is forecast to add 0.5°C to long-term warming, but this positive feedback is inherently a slow one because of the slow propagation of heat through ocean sediments.

The loss of major ice sheets is also a slow process, but it may already be underway. The Greenland ice sheet is thought to be a relic of the last ice age, which if removed could not regrow under the present climate. It is already losing mass and this may be irreversible thanks (in part) to a strong positive feedback whereby melting causes the ice sheet's surface altitude to drop, warming it up further and causing more melting. In Antarctica, the West Antarctic ice sheet and parts of the East Antarctic ice sheet are grounded on the sea floor well below sea level. Depending on the depth profile of the sea floor, the 'grounding line' of the ice sheet where it parts from the sea floor can retreat abruptly, dislodging armadas of icebergs into the ocean and increasing sea level. The shrinkage of major ice sheets is already a key contributor to sea-level rise together with glacier melt and the expansion of the ocean as it warms up. On a business-as-usual scenario, sea-level rise could increase to a metre this century and in the long term to tens of metres.

Marine ecosystems and biogeochemistry

The ocean is already acidifying because CO_2 reacts with seawater to form carbonic acid. This poses a threat to organisms that precipitate carbonate, including corals and many members of the plankton and benthos. Corals are also sensitive to ocean warming which can cause bleaching events. Hence, if we continue business as usual, large-scale loss of coral reefs has been forecast.

On millennial timescales, acidified waters will propagate into the deep ocean tending to dissolve calcium carbonate sediments there. This will release alkalinity, in turn allowing the ocean to take up more CO_2. At the same time, global warming and the acidification of soil waters by CO_2 will accelerate carbonate and silicate weathering on land tending to replenish the ocean with alkalinity. Over hundreds of thousands of years excess silicate weathering will remove the fossil carbon we add to the atmosphere and deposit it in new carbonate rocks. However, by then the next ice age will already have been prevented, and the Quaternary glacial–interglacial cycles could cease entirely.

Carbon dioxide is not the only human waste product with long-term consequences for the Earth system. The increased input of nitrogen and phosphorus to the land is already fuelling anoxic conditions in freshwaters and some shelf seas. If this continues for millennia (a big 'if') then it will significantly increase the nitrogen and phosphorus content of the ocean, risking triggering a global anoxic event, because anoxia enhances phosphorus recycling from coastal shelf sea sediments fuelling more productivity and anoxia—a potent positive feedback (Chapter 4). Ocean anoxia is further encouraged by warming of the ocean, which reduces the solubility of oxygen in the water and tends to stratify the ocean, isolating deeper deoxygenating waters from the atmosphere.

Emergent simplicity

As Niels Bohr put it, 'prediction is very difficult, especially about the future'. This is especially true for complex systems such as the Earth system, and it is why many careers-worth of effort have already gone into constructing Earth system models. Still, complex as it is, the Earth system can display some 'emergent simplicity'. For example, the linear relationship between cumulative CO_2 emitted and global temperature change appears robust across a range of models. We can thus project some of the consequences of our collective activities with confidence, even if we cannot predict how human societies are going to develop. Other properties of the Earth system, such as tipping points, remain more difficult to forecast. Still, through a combination of examining the past behaviour of the Earth system, understanding the processes at work, and incorporating this understanding in models, progress is being made. In the future we can imagine a new generation of Earth system models that allow us to examine 'planetary boundaries' other than just climate change—such as the limits on cumulative phosphorus and nitrogen addition to avoid widespread deoxygenation of the ocean. Perhaps we will even try to simulate human societies as an interactive part of the Earth system, if only to scope out the possible directions that lie ahead. In Chapter 7, we consider one of those directions—what it will take to achieve long-term sustainability.

Chapter 7
Sustainability

Whilst human transformation of the planet was initially unwitting, now we are increasingly collectively aware of it. This poses a challenge to Earth system science because we humans have conscious foresight and a sense of purpose that (as far as we know) has never been part of the Earth system before. This changes the Earth system fundamentally, because it means that one species can consciously, collectively shape the future trajectory of our planet. We know our current way of living is unsustainable, but we are still trying to work out what a sustainable and prosperous future looks like. This is an opportunity for Earth system science, because it is the field that can tell us what makes a sustainable Earth system and what does not. This chapter outlines how Earth system science can help humanity in our quest for sustainability, starting with the lessons we can learn from Earth history.

Earth history lessons

The Earth's biosphere is an outstanding example of a sustainable system. It has been thriving for over 3.5 billion years, initially as a world of prokaryotes, but now as a world supporting complex life. In that time, the Sun has got steadily brighter, large rocks have hit the planet, and the inner Earth has occasionally injected large pulses of molten material into the surface Earth system. Yet despite these perturbations, conditions at the surface of the Earth have not only remained habitable, but life has thrived. Yes, there have been some near fatal catastrophes like the snowball Earth events, or the end-Permian extinction, but these are the exceptions, not the rule. So, what are the secrets of long-term sustainability?

The first secret is to couple a sustainable energy supply to material recycling (Figure 26). The Earth system's primary energy source is sunlight, which the biosphere converts and stores as chemical energy. The energy-capture devices—photosynthesizing organisms—construct themselves out of carbon dioxide, nutrients, and a host of trace elements taken up from their surroundings. Inputs of these elements and compounds from the solid Earth system to the surface Earth system are modest. Some photosynthesizers have evolved to increase the inputs of the materials they need—for example, by fixing nitrogen from the atmosphere and selectively weathering phosphorus out of rocks. Even more importantly, other heterotrophic organisms have evolved that recycle the materials that the photosynthesizers need (often as a by-product of consuming some of the chemical energy originally captured in photosynthesis). This extraordinary recycling system is the primary mechanism by which the biosphere maintains a high level of energy capture (productivity).

The second secret of sustainability is self-regulation. To maintain stable, habitable conditions the Earth system must possess negative feedback mechanisms, such as the silicate weathering feedback that stabilizes temperature over the long term. These negative feedbacks give the Earth system resilience—meaning that if something hits the system, it tends to bounce back to its initial state. Resilience is literally a measure of how fast it bounces back. The role of life in some negative feedback mechanisms—such as the amplification of silicate weathering—increases the resilience of the Earth system. Of course there are positive as well as

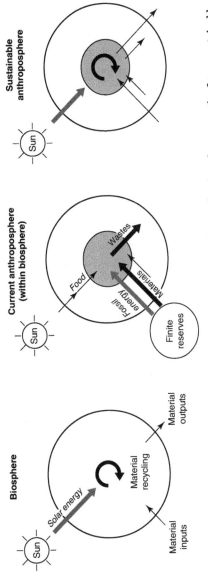

26. Energy and material flows in the biosphere, the human sphere (anthroposphere), and a prospective future sustainable anthroposphere.

negative feedbacks in the Earth system. However, its long-term stability tells us that overall, negative feedback has had the upper hand. Why this is the case is a subject of ongoing debate.

Whilst it is tempting to infer from the long history of life that the Earth system will be resilient to our activities as a species, this could be false confidence. Past survival is not necessarily a good guide to future stability. The reason is that our very existence requires the Earth system to have had a particular kind of history: one in which life survived and oxygen levels rose to the point that we could arise and observe such a history. This is an application of what cosmologists call the 'weak anthropic principle'. It means we should not be too surprised that the Earth system has exhibited predominantly negative feedback up to now. This is no guarantee that it will continue to do so. Conceivably something could arise within the Earth system that profoundly disrupts it—even to the point of killing off all life. Indeed, some of us suspect that 'something' could be us.

Exponential growth meets finite resources

At the heart of the sustainability challenge is a tension between positive and negative feedback. Exponential growth can come from a positive feedback that is inherent to biology—life begets more life. But exponential growth is always ultimately constrained by finite resources, which impose a negative feedback on growth. This idea that exponential growth will eventually be constrained by finite resources dates at least as far back as Thomas Malthus' 1798 'An Essay on the Principle of Population'. It was fundamental to Darwin's formulation of the theory of natural selection: when resources are constrained, population numbers stabilize, competition for resources ensues, and the 'fittest' organisms (the ones that leave the most descendants) come to dominate the world.

The principle took on a broader form in the 1972 book, *The Limits to Growth*, by Donella H. Meadows, Dennis L. Meadows, Jørgen

Randers, and William W. Behrens III, which examined the interaction between exponential growth of human activities and finite resources using an early global system model called World3. The authors modelled five interacting variables: world population, industrialization, pollution, food production, and resource depletion. Two model scenarios led to overshoot and collapse this century, and a third produced a stabilized world. The work was widely critiqued by economists, who argued that actively constraining resource consumption would seriously compromise ongoing improvements in human well-being. Out of this argument emerged the grand compromise of 'sustainable development': that we must simultaneously strive for human betterment and for environmental sustainability.

Perhaps the clearest positive link between human development and sustainability is that it generally leads us to have fewer children. This means that human population is destined to stabilize. Indeed fertility rates have already fallen below the replacement level in many developed countries. Hence if development is realized globally, we can project a declining human population in the long term. However, development also increases energy and material consumption, which have become decoupled from population growth, and continue to grow in an exponential fashion. Thus, the fact that population growth has been slowing down since the 1960s is not stabilizing our collective impact on the planet. This means that the sustainability challenge is not primarily about stabilizing population (although that will help); it is about changing our sources of energy and what we do with materials.

Sustainable energy

The field of 'industrial metabolism' (or 'industrial ecology') views human societies as having intertwined flows of energy and materials, just as organisms, ecosystems, and the Earth system do. Whilst the biosphere has achieved occasional jumps in energy input in the past, for most of its history energy inputs have been stable. Up to now humans have increased energy input to the biosphere by about a tenth, most of that occurring since the 'great acceleration'—with world energy consumption growing from around 100 EJ/year in 1950 to roughly 500 EJ/year in 2010. Future projections suggest energy demand could rise to over 1,000 EJ/year by 2050. Growing energy demand cannot be met indefinitely—but we aren't anywhere near the limits yet.

At present our major energy source is embodied in materials— fossil fuels. Whilst we continue to burn fossil fuels, we know we are living on borrowed energy. Fossil fuels are a finite resource from the Earth's crust. Hence negative feedback will limit the growth in their consumption. Oil extraction is destined to peak first, followed by gas, then (eventually) coal. However (as we saw in Chapter 6), there is enough fossil fuel—primarily coal—to warm the Earth's climate by the order of 10°C. This could trigger a different kind of negative feedback on growth from its profoundly damaging effects. We could instead take the option of burning fossil fuels and capturing and storing the CO_2 given off—with an attendant energy penalty—but that cannot be described as a 'sustainable' energy source.

Nuclear fission power relies on finite supplies of fissionable material so it is not indefinitely sustainable either. Nevertheless, the future potential for energy generation from fissionable uranium and thorium is greater than that from fossil fuels. Nuclear fusion is what powers the Sun and all stars, and if it can be tamed then supplies of 'fusionable' material are orders of magnitude greater still. Nevertheless, the Sun looks set to continue being the dominant source of energy for the biosphere in the long-term future.

Human intervention could greatly increase the fraction of sunlight that enters the biosphere, because solar energy is not very efficiently converted by photosynthesis, which is 1–2 per cent efficient at best. Efforts are underway to improve the efficiency of photosynthesis, which could benefit production of both food and biofuels. However, there remains a large efficiency gap between photosynthesis and other means of solar energy capture. Solar photovoltaic (PV) panels are typically ~20 per cent efficient in converting sunlight to electricity, and recently a combined PV and solar thermal capture device with 80 per cent efficiency has been produced. The amount of sunlight reaching the Earth's land surface (2.5×10^{16} W) dwarfs current total human power consumption (1.5×10^{13} W) by more than a factor of a thousand. Hence solar power could support significant future growth in human energy consumption, but not indefinite growth—because there is only so much sunlight hitting the Earth.

This would of course mean devoting a fraction of the Earth's surface to solar energy capture, and solving some challenges of energy transfer and storage. Solar and most other renewable energy sources are intermittent. Hence they require energy storage and/or a global electricity supergrid that can transfer power from where the Sun is shining to where it is needed. Energy can be stored in many forms. The biosphere stores solar energy in chemical form and the human equivalent of this could be hydrogen fuel, or synthetic hydrocarbons.

Material recycling

Currently we are mining a range of finite resources from the Earth's crust, including phosphorus for fertilizer, iron, aluminium, and a host of trace metals (as well as fossil fuels). We are also fixing a large amount of nitrogen from the atmosphere. We use these materials in our industrial economy or to grow food. Then we typically dump the waste products either back in the land, release them to the atmosphere, or let them run off into freshwaters and the ocean. This in turn generates environmental changes. For long-term sustainability there will have to be a fundamental shift towards much greater recycling, because resources are finite and because accumulating waste products are causing escalating environmental damages.

Recycling in turn requires energy. In the biosphere, recycling is fuelled by the chemical energy captured in photosynthesis being consumed in heterotrophic reactions. For example, soil organisms get their energy from respiring organic matter with oxygen and in the process liberate nutrients back into the soil (as well as returning CO_2 to the atmosphere). This is a neat trick that supports life directly with energy, and indirectly through recycling resources. We cannot quite pull off that trick with most of the materials of civilization, because they don't embody energy in the same way. However, we could create an industrial ecology in which solar energy fuels material recycling.

Although the total amount of many elements in the crust is vast, as we extract concentrations of an ever-diminishing purity the energy required to mine and refine them increases. This should put an incentive on recycling what we already have in circulation in the surface Earth system, because it is more energy efficient. For example, recycling iron (and steel) requires only about a quarter of the energy of refining it from iron ore. Sure enough, in North America, where the highest grade ores have long since been mined, around half of the steel used in new construction is recycled. Of course, if the total amount of an element in use in societies is increasing—as it still is for iron globally—then there has to be an input (in this case from the crust). But with population stabilizing we can anticipate a stabilization of the total amount of built infrastructure.

Some materials are basic to our survival, such as the phosphorus and nitrogen in our food, much of which derives from fertilizer application. The total amount of nitrogen in the atmosphere is so vast it will never be limiting, as long as we have a sustainable source of energy to fix it. However, there has been recent concern that rock phosphate reserves could become limiting and we might be approaching 'peak phosphorus' production. This means going to less concentrated or harder to get to reserves and this pushes the price up. Whether this peak is near or far, we clearly need to be thinking about much more efficient usage and recycling of phosphorus.

Planetary boundaries

The accumulating material waste products of our current industrial metabolism have the potential to transgress the limits to the healthy functioning of the Earth system. The concept of planetary boundaries is an attempt to define these limits and a corresponding 'safe operating space for humanity' (Figure 27). Beyond the boundaries the Earth system would be forced out of a stable 'Holocene-like' state. A total of nine boundaries have been proposed covering climate change, ocean acidification, ozone depletion, biogeochemical flows, freshwater use, land-use change, biodiversity loss ('biosphere integrity'), atmospheric aerosols, and novel entities including chemical pollution. Numerical values have been suggested for some of these boundaries—set at the lower end of a range of uncertainty, in order to remain within a 'safe operating space'. For example, the climate change boundary was set at 350 ppm CO_2—a level already transgressed.

Whilst there is much debate over the precise setting of particular planetary boundaries, the fundamental notion that there are limits to the safe functioning of the Earth system is less contested. Some of these limits can be transgressed before resource constraints kick in. For example, we clearly have the technological capacity to destroy the ozone layer. Furthermore, available fossil fuel—primarily coal—gives us the capacity to take the Earth system well outside of the Holocene (or even Quaternary) state.

We obviously need to consume a certain amount of energy and resources to sustain human well-being. For materials such as phosphorus, there is a tension between our basic need to feed the human population, its finite reserves, and the environmental

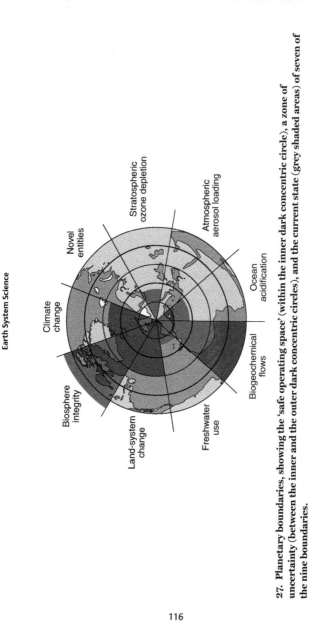

27. Planetary boundaries, showing the 'safe operating space' (within the inner dark concentric circle), a zone of uncertainty (between the inner and the outer dark concentric circles), and the current state (grey shaded areas) of seven of the nine boundaries.

consequences of its use. There are also potential tensions and trade-offs between planetary boundaries. For example, if we choose to use less phosphorus and nitrogen in agriculture and this reduces crop yields, then more land use may be required to feed the human population, with attendant effects on biodiversity. Accepting that there are planetary boundaries and trade-offs between them, we need consciously to design sustainable ways of operating within them.

A self-aware feedback system

Human consciousness is—as far as we know—a fundamentally new property of the Earth system. When combined with our technological capability to change the world, it brings the potential for a new type of feedback control in the Earth system. Up to now all the Earth system's regulatory (and destabilizing) feedbacks have arisen and functioned unconsciously. Looking ahead there is the possibility that we will introduce conscious, purposive—'teleological'—feedback control into the Earth system (Figure 28). In a sense we are already doing this.

Whilst the champions of the Industrial Revolution did not consciously set out to alter the climate, we cannot pretend that we don't now know the climatic consequences of our ongoing industrial activities. Indeed, when Arrhenius published his calculation of the global warming effect of CO_2 in 1896 he recognized that industrial activities were adding CO_2 to the atmosphere and warming the climate, which he saw then as a good thing. Now we think otherwise, and the effort to mitigate fossil fuel CO_2 emissions, limited though it has been thus far, represents a conscious negative feedback activity. It is an attempt to reduce the extent of expected negative impacts of climate change to stay within that 'planetary boundary'.

Monitoring is essential to any teleological feedback control system. If we do not know the state of the Earth system then we cannot measure our activities towards or away from any goals we

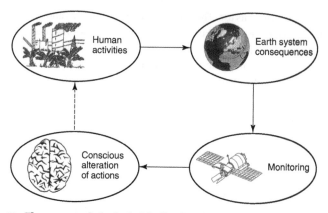

28. The concept of teleological feedback within the Earth system.

set. Up until now Earth system monitoring has been largely instigated out of curiosity and a desire to record the consequences of our actions. Currently we are monitoring the Earth system at a range of levels, from the surface to space, in increasing spatial and temporal detail. There are still massive gaps in spatial observation—for example in monitoring the state of the carbon cycle, of biodiversity, or of the deep ocean. There are also gaps in temporal observations, especially reconstructing them before the modern era. However, progress is being made, for example in monitoring the Atlantic Ocean's overturning circulation and reconstructing its behaviour in the past.

Early warning signals

As well as recording steady changes in the Earth system, such observations can provide an important clue as to whether abrupt changes are approaching. The idea is a simple one: a system that is losing stability becomes progressively more sensitive to perturbations. This means it takes longer to recover from any given perturbation (a loss of resilience), and also it tends to move further under a given perturbation (an increase in variability). We don't need to deliberately perturb a system to observe these signals—the Earth system's own internal variability acts as a continual source of 'noise' to which we can observe responses. If we see parts of the Earth system becoming more sluggish in response to natural fluctuations then this indicates they are losing resilience.

Such early warning signals have been found prior to some past abrupt climate changes and in model simulations being forced towards tipping points, such as a collapse of the Atlantic overturning circulation. In reality they could give us a useful clue as to which parts of the Earth system are becoming less resilient, and thus could provide a cue for intervention action.

Response options

Several response options can be envisaged to an early warning that part of the Earth system is losing stability, or that we are approaching a particular planetary boundary. Mitigating the root cause of change is the most obvious and common-sense strategy. It has been successfully employed to reduce emissions of ozone-depleting chlorofluorocarbons (CFCs) globally, and to tackle air pollution and nutrient loading of freshwaters and coastal seas—at least regionally. However, we have yet to make much headway with global fossil fuel CO_2 emissions, because they underpin our current economic system. Still, other intervention actions are conceivable.

One response to observing a loss in resilience of part of the Earth system could be consciously to intervene to try and strengthen negative feedbacks to maintain a desired state. An example of this would be if we observed weakening of the land carbon sink and intervened to try to strengthen it. This could take the form of stopping deforestation and deliberately afforesting. It could involve reducing the tillage of agricultural lands to help them store soil carbon, or converting biomass to charcoal and adding it to soils as long-lived 'biochar'. It could also involve adding crushed silicate minerals to soils in a deliberate effort to enhance weathering. These methods can bring local benefits to ecosystem resilience, such as reducing soil erosion, enhancing water and nutrient retention, and counteracting acidification. Other methods of deliberate carbon dioxide removal from the atmosphere have also been suggested, such as direct air capture. All of these methods would have to be monitored to verify their effects on the carbon cycle.

The bolder corollary of building resilience in the Earth system (i.e. strengthening negative feedbacks) is the idea of deliberately strengthening positive feedbacks in order to propel the transition from an undesired state into a desired state. Examples of this are proposals to green parts of the Sahel and the Sahara, or the Australian desert. Observations suggest that these regions have exhibited a greener and wetter state in the past, which was self-maintaining thanks to the vegetation cover. A 'great green wall' project is already envisaged to halt desertification in the Sahel by planting a band of trees. Augmenting this with supplies of desalinated seawater and nutrients might help enable the transition to a stable, green Sahara state.

Even larger scale deliberate interventions in the climate system are also being discussed under the banner of 'geoengineering' (or 'climate engineering'). In particular, 'solar radiation management' methods involve trying to reflect more sunlight back into space to cool the planet. A much-discussed way of doing this is to inject sulphate aerosols (tiny particles) into the stratosphere to mimic the cooling effect of past volcanic eruptions. This is a potent technology because only a modest mass of aerosol would be required to cancel out the warming effect of the vast mass of CO_2 we are adding to the atmosphere.

However, such aerosols are short-lived, so would have to be continually replenished for many centuries to avoid rapidly unmasking the underlying enhanced greenhouse effect. At the outset, we don't know precisely how much aerosol is needed to produce a given level of cooling, nor do we accurately know the side effects on other aspects of the Earth system, such as precipitation patterns. Hence any such geoengineering scheme would have to monitor consciously its effects and adjust the intervention accordingly—and that would be an example of teleological feedback.

Such proposals may create a greater risk than the one they are trying to avoid. Nevertheless, they serve to illustrate that we may be on the cusp of consciously steering the future trajectory of the Earth system. Indeed we probably already have, just in a different context. As Oliver Morton has highlighted, the 'green revolution' was a consciously designed effort to enrich large parts of the global land surface with nutrients to support a rapidly growing human population. It has also had the (perhaps unexpected) side benefit of protecting large areas of land from the plough.

Earth system economics

Most activities towards long-term sustainability require cooperation and are altruistic towards other members of our species (and the biosphere). The idea of global teleological feedback goes a stage further and would effectively represent a new level of biological organization, able to regulate the global environment. However, evolutionary theory tells us that cooperation is notoriously unstable, because it is prone to 'cheating' by free-riders who enjoy the benefits without contributing to the costs. Earth history confirms that the successful emergence of new levels of biological organization—such as the eukaryote cell, or social groups of animals—is rare. Human history, however, shows an unusual amount of group-level selection, and recent social progress can be seen as a conscious effort to free ourselves from the sometimes brutal constraints of natural selection.

The tension between individuals maximizing their short-term personal gains, and what is best for the system as a whole in the long-term, is encapsulated in the 'tragedy of the commons'. In a nutshell, as individuals we are all trying to increase our personal well-being, whereas the environment is a resource that we all share. Hence anything we do to improve (or degrade) the environment will be shared by everyone, including those not making an effort themselves. The 'tragedy' is that the division of costs and benefits is unequal so that the rational course of short-term individual action tends to be to degrade the shared environment—even if this is worse for everyone in the long term. Such a 'tragedy' is now enfolding in the global commons—the atmosphere and oceans. Happily there are several ways out of the 'tragedy', for example collective regulation that puts a price on polluting the environment. In the case of climate change this would be a price on CO_2 and other greenhouse gas emissions (and a corresponding reward for removing these gases).

Such a shift in pricing does not fundamentally alter the model of economic growth, it just seeks to steer it in a different direction. Already there is evidence of economic growth decoupling from the accumulation of material waste products, through the shift towards an economy of information exchange rather than material exchange. This might be further encouraged by the introduction of deliberate regulatory instruments that make the recycling of materials economically advantageous. Storing and exchanging information, however, still requires energy and materials, and efficiency gains are fundamentally limited by the second law of thermodynamics. Therefore a complete decoupling of economics from the Earth system is impossible. As Adam Smith recognized, this means that economic growth cannot continue indefinitely. An outstanding task therefore is to formulate a steady state 'Earth system economics' that supports long-term human and planetary well-being.

Broadening the field

If we consider ourselves and our societies as integral parts of the Earth system, and we take seriously the new properties that humans bring to the Earth system, then this requires a new kind of Earth system science. It has to integrate elements of the social sciences at least insofar as they help us to understand the role of human agency in planetary functioning. This could change the nature of Earth system models and the ways in which we use them. Instead of making predictions based on some set of assumptions about future human activities—as if we lived outside of the system—human activities and agency could become a more integral part of the models. Equally, Earth system considerations call for some rethinking of economics and a wider social discussion about what kind of future we want, which will engage the arts and the humanities as well as the social sciences.

Chapter 8
Generalization

This book has introduced how one habitable planet—the Earth—can be studied as a system. However, in just the past few years, scientists have made the remarkable discovery that there are potentially habitable planets orbiting other stars. Just as humanity's first view of the Earth from space changed how we saw and studied our home planet, our first 'view' of an Earth-like planet around another star will surely change our perspective again. This concluding chapter explores how our understanding of the Earth system can be generalized into a science of habitable worlds in general.

Lifespan of the biosphere

If humans manage to find a sustainable relationship with the Earth system then we might expect to persist for a million years or so—the typical lifetime of a mammal species. If we are lucky (or very smart) we might stretch this to ten million years. Complex life can expect a longer 'lifetime' on Earth and prokaryote life an even longer one, but not an indefinite one.

The problem is that like all stars on the 'main sequence' (which generate energy through the nuclear fusion of hydrogen into helium), the Sun is burning inexorably brighter with time—roughly 1 per cent brighter every 100 million years—and eventually this will overheat the planet. This is compounded by a positive feedback—heating evaporates water into the atmosphere, which traps more heat. This is already the strongest positive feedback in the climate system and it is destined to get stronger as the warming atmosphere gets more saturated with water vapour, making it more opaque to the heat radiation coming off the Earth. Eventually this will cause a 'runaway greenhouse effect' in which the Earth is unable to shed heat to space as fast as it is coming in. Then as the temperature escalates, the oceans will evaporate.

Before runaway happens, a fatal 'moist greenhouse' will occur, in which the atmosphere becomes like a steam pressure cooker, raising the boiling point of the remaining oceans, and expanding the lower atmosphere. Water molecules will reach the upper atmosphere and be split apart by the intense radiation there, allowing the hydrogen they contain to be lost to space and desiccating the planet. Before all water is lost, temperatures will have become too hot for complex life. A rough upper tolerance limit for eukaryotes is 50°C, whereas some 'extremophile' prokaryote life forms can tolerate over 100°C (if extra pressure has raised the boiling point of water above this). So the Earth system will probably revert to a world of archaea and bacteria before it desiccates.

Can this ultimate fate for life be delayed? Over Earth history, the silicate weathering negative feedback mechanism has counteracted the steady brightening of the Sun by removing carbon dioxide from the atmosphere. However, this cooling mechanism is near the limits of its operation, because CO_2 has fallen to limiting levels for the majority of plants, which are key amplifiers of silicate weathering. Although a subset of plants have evolved which can photosynthesize down to lower CO_2 levels, they cannot draw CO_2 down lower than about 10 ppm. This means there is a second possible fate for life—running out of CO_2. Early models projected either CO_2 starvation or overheating (Figure 29)

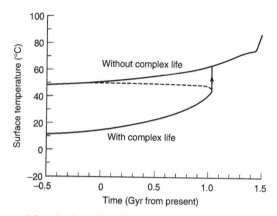

29. A model projection of the lifespan of the biosphere in which complex life overheats and the Earth system switches to a hotter steady state (without biological enhancement of silicate weathering), followed later by a moist greenhouse.

occurring about a billion years in the future. Whilst this sounds comfortingly distant, it represents a much shorter future lifespan for the Earth's biosphere than its past history. Earth's biosphere is entering its old age.

The habitable zone

There is another way to think about the limits to planetary habitability and that is in space rather than time. The habitable zone (Figure 30) represents the range of distances around a star (or equivalently the range of stellar luminosities) at which a terrestrial, rocky planet can support liquid water at its surface. If a planet is too close to its parent star, then overheating and loss of water will occur, marking the inner edge of the habitable zone. If a planet is too far away from its parent star, then overcooling and freezing out of water will occur, marking the outer edge of the habitable zone. This is the spatial equivalent of the 'faint young Sun puzzle' that we met in Chapter 1.

The outer edge of habitability is defined where the build-up of CO_2 can no longer keep a planet above freezing temperature. Instead, at very high concentrations, CO_2 becomes a net cooling agent because it scatters more incoming solar radiation than it traps heat radiation coming off a planet. This effect is known as 'Rayleigh scattering' and happens with all small gas molecules (Rayleigh scattering by the dominant gases N_2 and O_2 in the atmosphere today is what makes the sky appear blue and the Sun yellow). Ultimately Rayleigh scattering will make it impossible to escape from a snowball state by the build up of CO_2.

Habitable zones exist around all stars on the main sequence which burn steadily brighter over time, thus causing their habitable zones to move steadily outwards. Stars differ in mass and therefore luminosity, affecting the location of their habitable zones, but these factors can be readily accounted for in models. The classic model used to estimate the boundaries of the habitable zone was developed by Jim Kasting and colleagues. It assumes an Earth-like planet in terms of mass and atmospheric pressure that is tectonically active and has a water cycle. Hence the silicate weathering negative feedback is assumed to operate and this broadens the width of the habitable zone, counteracting decreasing luminosity (i.e. moving further away from a star) by increasing atmospheric CO_2, and counteracting increasing luminosity (i.e. moving towards a star) by removing CO_2. Without this feedback adjusting the CO_2 level, the habitable zone would be much narrower.

The latest estimates from Kasting's group put the outer edge of the Sun's habitable zone currently beyond the orbit of Mars (Figure 30) at 35 per cent of the luminosity received by the Earth today. This is consistent with evidence that liquid water has flowed on the surface of Mars in the past. However, under a fainter young Sun, early Mars is predicted to have been beyond the outer edge of the habitable zone, and its small size (10 per cent

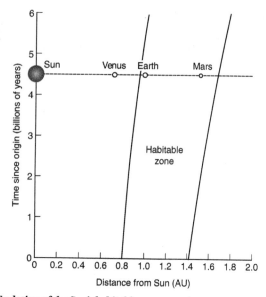

30. Evolution of the Sun's habitable zone over time.

of the Earth's mass, with correspondingly weak gravity and an irregular magnetic field) means that it has now lost most of its atmosphere and water to space. Venus is estimated to have always been inside the inner edge of the habitable zone, consistent with it having undergone a runaway greenhouse effect. In fact, the inner edge of the habitable zone is estimated to now be perilously close to the Earth, requiring only a 1 per cent or 3 per cent increase in solar luminosity to trigger a moist or a runaway greenhouse fate for the Earth. This implies only a 100–300 million-year future for Earth's biosphere, but should be viewed as a minimum estimate because it comes from a simple model that ignores circulation of the atmosphere and changes in cloud cover. Such a conservative model provides a good starting point for estimating the habitable zones around other stars and thus helping guide the search for potentially habitable extrasolar planets.

Extrasolar planets

In just the last two decades thousands of planets have been discovered orbiting other stars. At the time of writing over 1,500 'exoplanets' have been confirmed and over 3,000 more 'candidates' have been detected, most of them by the Kepler space observatory, which detects planets if they 'transit' between us and their parent star, causing a dip in the starlight received. (Fewer than 1 per cent of Earth-like planetary orbits are expected to pass through the line of sight between us and a star, hence over 190,000 stars were monitored by Kepler.) This large sample size has given us some idea of what a typical planet and a typical planetary system looks like. The average star has at least one planet. The most common type of planet has a radius between one and four times that of the Earth—that is, up to the radius of Neptune—making our Solar System an anomaly, because it has no planets in this intermediate size range. If we consider 'super-Earths' in the range of between one and two times the Earth's radius, these are sufficiently common that roughly 10 per cent of Sun-like G-type stars are estimated to possess one in the habitable zone. The proportion rises to 40–50 per cent for the cooler, fainter M- and K-type stars observed by Kepler. So the next time you look into the night sky, consider the remarkable thought that at least one in every ten stars you see could have a neighbouring planet harbouring liquid water at its surface.

As of late 2014, the closest candidate for an Earth twin is the rather innocuously named Kepler-186f—a planet with roughly 1.1 times the Earth's radius orbiting a typical M-type star. Kepler-186f is the outermost of five planets, orbiting its star at about 40 per cent of the distance of the Earth from the Sun. As its star is cooler and fainter than the Sun, this puts Kepler-186f well within its star's habitable zone, actually towards the outer edge, in a 'Mars-like' position (Figure 31). If Kepler-186f has a CO_2-rich

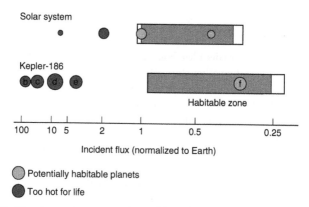

Solar system

Kepler-186

b c d e

f

Habitable zone

| 100 | 10 | 5 | 2 | 1 | 0.5 | 0.25 |

Incident flux (normalized to Earth)

🔘 Potentially habitable planets

⚫ Too hot for life

31. Planetary system comparison of the Solar System (with Earth and Mars in the habitable zone) and the system of planets around the star Kepler-186 (with the planet Kepler-186f in the habitable zone).

atmosphere and water, it is likely that some of that water is in liquid form at the surface.

Abundant carbon dioxide and water are not guaranteed on a planet because these are amongst the 'volatile' compounds that are not stable at the high temperatures close to a young star when rocky planets first form. They can be trapped in the mantle and subsequently out-gassed or brought to a planet by collisions with meteorites and comets originating further from their parent star. However, simulations of solar system formation suggest that the supply of volatiles from collisions is a haphazard affair. Thus whilst all exoplanets in the habitable zone should receive some water and CO_2, the amount is likely to vary considerably—and with it their habitability.

Life detection

The discovery that up to half the stars could have a potentially habitable planet orbiting them considerably increases the odds that there are other planets harbouring life near to us in our

Earth System Science

130

galaxy. Of course those odds also depend on the ease or difficulty of life evolving. However, the relatively rapid appearance of life on Earth after the 'late heavy bombardment' already hints that the origin of life is not that difficult. The question is: can we detect life on an exoplanet if it is there?

We cannot realistically go to an exoplanet. The nearest habitable one is estimated to be around twelve light years away. With a plausible escape velocity from the Sun of up to 100 km/s it would take 36,000 years to get a spacecraft there. Then communication would be rather stilted with a twelve-year delay in each direction. Instead we need a remote means of life detection.

This brings us full circle, back to Jim Lovelock's original idea to detect life through its effect on the composition of a planet's atmosphere (Chapter 1). Remember that Lovelock suggested that the signature of life is extreme disequilibrium in the mixture of atmospheric gases. On Earth today the coexistence of methane (CH_4) and oxygen (O_2) in the atmosphere is an example of such extreme disequilibrium. On exoplanets other disequilibrium mixtures are conceivable, some of which, like sulphur dioxide (SO_2) and hydrogen sulphide (H_2S), might be maintained by volcanic out-gassing. So, the search needs to be restricted to clearly biogenic disequilibrium pairs of gases like oxygen and methane.

The approach to detecting them is to use a spectrometer to look for the characteristic absorption bands (wavelengths) of different gases in the spectrum of radiation coming from an exoplanet. Some gases have more absorption bands than others, and absorption in each band depends on the concentration of gas. This poses a problem because extreme disequilibrium means that two reactants want to react together very rapidly, tending to reduce their concentrations. In the case of Earth's atmosphere today this puts methane at a concentration of around one part in a million, which means its modest absorption bands would not be detectable if we were observing Earth from another star. Oxygen and its product ozone, on the other hand, would be detectable.

Around stars that emit less UV radiation than the Sun, atmospheric chemical reactions would slow down considerably, allowing gas concentrations to build up and perhaps permitting the detection of methane in the presence of oxygen—or other disequilibrium pairs of gases. However, researchers are currently shifting their attention to the detection of single 'biosignature' gases. For gases like oxygen that can be made abiotically, this suffers the problem of false positives. An alternative is to look for gases made only by life, such as dimethyl sulphide, isoprene, or methyl chloride, but these tend to be at much lower concentrations.

Despite the caveats, the prospect of finding another inhabited planet is no longer science fiction. It could happen in the next decade. It is of course a phenomenal technological challenge, requiring expensive space-based telescopes, but scheduled missions are moving in the right direction. Set to launch in 2017, the Transiting Exoplanet Survey Satellite will search the entire sky for rocky planets around nearby bright stars. This should provide candidate 'exo-Earths' in the habitable zone of their stars that are much nearer and therefore brighter than those found by Kepler. Then the James Webb Space Telescope, due to launch in 2018, will have the resolution necessary to begin to characterize their atmospheres as they transit their parent stars. If life is common and we are lucky, it could detect alien life. If not, a further generation of space telescope that 'subtracts' or shades out the light of parent stars will greatly increase the prospects for resolving atmospheric composition.

Exoclimates

The recent exoplanet discoveries and the prospects for remote life detection make it an exciting time to start formulating a science of habitable planets in general, not just the Earth in particular.

Researchers have started by generalizing three-dimensional models of the Earth's climate and using them to revise estimates of the habitable zone. This is helped by the principles of atmosphere and ocean circulation being the same everywhere, involving the Navier–Stokes equations of fluid dynamics, gravity, and the Coriolis force, which depends on a planet's rotation rate. Initial studies have shown that atmospheric circulation effects could greatly extend the inner edge of the habitable zone towards a star—delaying the runaway greenhouse effect. The key reason is that in a three-dimensional, circulating atmosphere there should still be some regions of dry descending air, and through these drier parts of the atmosphere, heat radiation can escape more effectively to space—preventing runaway.

If planets with slower rotation rates than the Earth are considered, this could extend the habitable zone even further inwards. Simulations suggest that deep cloud cover would develop on the day side of a slowly rotating planet, reflecting much of the light reaching it from its parent star back to space. Heat transport through the atmosphere would also be very effective, with giant convection cells stretching from the equator to the poles and from the day to the night side of the planet. These cooling effects could extend the inner edge of the habitable zone to twice the stellar flux experienced by the Earth, thus including planets that are sufficiently close to their parent star that they are 'tidally locked' to it, with the same side of the planet always facing the star.

The outer edge of the habitable zone appears less sensitive to resolving atmospheric circulation and cloud cover. However, complex models do find that a frozen 'snowball' state is harder to enter and easier to leave than suggested by simple models. In particular, clouds over ice cover can help to keep the climate warm, because they are no more reflective than the ice underneath them, yet they trap heat radiation coming up from a planet's surface.

Exogeology

Existing assessments of the habitable zone assume that the CO_2 content of a planet's atmosphere can automatically adjust—thanks to the silicate weathering negative feedback—such that it is very low at the inner edge of the habitable zone and very high at the outer edge. This assumes a planet is tectonically active with some exposed land surface, but both assumptions can be questioned.

Plate tectonics is not guaranteed. Current theories suggest plate tectonics is more likely on a larger rocky planet, and is encouraged by the presence of liquid water. Mars lacks plate tectonics consistent with its small size, but so too does Venus, despite being only slightly smaller than the Earth. In fact Earth is predicted to be close to the lower size limit for plate tectonics and may have water to thank for its presence, and plate tectonics may in turn have been required to maintain liquid water in the face of a brightening Sun. On smaller exoplanets without plate tectonics, there would be no obvious mechanism to recycle carbon deposited in sediments back to the atmosphere. Instead volcanic activity would only inject small amounts of CO_2 from the planet's mantle and this would be readily mopped up by silicate weathering, leading to a persistently very low atmospheric CO_2 level, and restricting the outer edge of the habitable zone.

On the more common 'super-Earths', plate tectonics is a safer assumption, but the presence of exposed land surface is not guaranteed. This is because the stronger gravity of a super-Earth will dampen surface elevation, potentially leading to a 'water world' with nothing sticking above the ocean surface. With no exposed surfaces to weather and therefore no continental silicate weathering, CO_2 would build up in the atmosphere. However, there is a second removal process for CO_2, through reaction with the sea floor (basalt) as it is formed at mid-ocean ridges. This removal process is also thought to be temperature sensitive, so it could act to stabilize the CO_2 level and the climate—albeit in a hotter state. Furthermore, the higher sea floor pressure on a super-Earth is expected to cause more water to be removed to the mantle, reducing the ocean volume and allowing some land mass to be exposed and the silicate weathering feedback to operate.

Exobiospheres

Having considered the geological requirements for an exoplanet to be habitable, is there anything general we can say about the type of biosphere it could host?

A key consideration for any biosphere is its supply of energy. For an exoplanet in the habitable zone, its main source of energy will be the neighbouring star. Hence to achieve global significance, an exobiosphere needs to be fuelled by some form of photosynthesis, which transfers the energy from photons to electrons. The majority of stars are fainter than the Sun, which means that the individual photons of light they give off are generally lower in energy. Hence more photons would need to be captured to do any particular kind of photosynthesis. On Earth, anoxygenic photosynthesis requires one photon per electron, whereas oxygenic photosynthesis requires two photons per electron. On Earth it took up to a billion years to evolve oxygenic photosynthesis, based on two photosystems that had already evolved independently in different types of anoxygenic photosynthesis. Around a fainter K- or M-type star (orange to red in colour), oxygenic photosynthesis is estimated to require three or more photons per electron—and a corresponding number of photosystems—making it harder to evolve. Hence, as oxygen-breathing animals we should not be too surprised to find ourselves orbiting an unusually bright G-type star (white to yellow in colour)—it may be too hard to evolve oxygenic photosynthesis around fainter, more typical stars that give off lower energy photons. However, fainter stars spend longer on the main sequence, giving more time for evolution to occur.

Given these considerations, anoxygenic photosynthesis is likely to be more common in the universe than oxygenic photosynthesis. However, on an exoplanet the potential electron donors for anoxygenic photosynthesis—e.g. H_2, H_2S, and Fe^{2+}—are expected to be scarcer than water, just as they are on the Earth. This will limit the productivity of anoxygenic biospheres, giving them less energy with which to alter the composition of their atmospheres, and making them harder to detect from afar. Oxygenic photosynthesis on the other hand supercharges a biosphere with energy, giving it more potential to transform its atmosphere and be detectable from afar. Thus, although oxygenic biospheres are probably rarer than anoxygenic ones, they should be easier to detect. The best place to look for them may be around hotter stars like our own Sun, with higher energy photons, and these are indeed the target of forthcoming space telescopes.

However, the biosignature of an oxygenic biosphere may not be the obvious one of oxygen and ozone in the atmosphere. As Earth history shows, the build-up of oxygen in an atmosphere requires elevated rates of hydrogen escape to space, which in Earth's case was probably fuelled by high concentrations of biogenic methane. Hydrogen escape also depends on planet mass and is expected to get more difficult on super-Earths. Also, even after the Great Oxidation on Earth, oxygen remained at modest levels and may have required the evolution of complex land life to increase it to modern levels, by enhancing the weathering of phosphorus from rocks.

Exo-Gaias?

Existing studies of the habitable zone tend to assume it is independent of the presence or absence of life. But a planet's habitability is likely to depend on whether it is inhabited. Recall

Lovelock's Gaia hypothesis that the presence of life increases the habitability of the Earth. Whilst some Earth system scientists disagree over the sign of the effect, most would agree that life affects Earth's habitability. By extension, the presence of an exobiosphere that is able to alter the atmospheric composition of its host planet in a detectable fashion is also likely to be altering its planet's habitability. But can we say anything general about how life affects habitability?

Let's start with the established mechanism to expand the habitable zone—the silicate weathering negative feedback. On Earth we know that silicate weathering is accelerated by the presence of land life, as a result of its quest for rock-bound nutrients. In fact the very low CO_2 level on the present Earth is in large part due to the biological amplification of silicate weathering. Without life, today's Earth would be hotter and some simulations suggest it could already have become uninhabitable for complex life—consistent with the Gaia hypothesis.

Other biospheres can also be expected to lower their atmospheric CO_2 level. First, if photosynthesis evolves, this tends to transfer CO_2 from the atmosphere to the crustal rocks of a planet in the form of dead organic matter. Second, enhanced silicate rock weathering is likely to be a generic solution to the common problem of scarcity of phosphorus at a planet's surface. The resulting lowering of CO_2 will tend to move the habitable zone of an inhabited planet closer to its star, extending the inner edge, but posing a cooling problem for biospheres that start life towards the outer edge of their habitable zone. We can also expect biospheres to evolve some fairly generic warming effects. Notably, recycling of organic carbon as methane is a simple and very ancient metabolism on Earth. Earth's early biosphere recycled up to half of the carbon captured in photosynthesis as methane, and methane is a more potent warming agent than CO_2. Hence the net effect is estimated to have been warming of the early Earth (although there is a potential problem that if the CH_4:CO_2 ratio approaches one then atmospheric haze is created that scatters sunlight back to space and cools a planet).

There are in fact many conceivable ways in which a biosphere can warm or cool its host planet. Life can produce other potent greenhouse gases, including nitrous oxide and carbonyl sulphide—or it can generate other cooling effects, such as the production of dimethyl sulphide increasing the albedo of clouds. Any effect on the climate will in turn generate a feedback loop, because biological processes are almost universally sensitive to temperature. The resulting systems of multiple feedback loops begin to sound too complex to say something general about their properties. But there are some simple feedback principles that can be applied.

If some life forms are making their planet more habitable this will tend to be self-reinforcing—positive feedback will encourage the spread of life. Conversely, if some life forms start to drive their planet towards the limits of their own habitability then this will be self-limiting—negative feedback will kick in to restrict the spread of life. Whether it can kick in quickly and strongly enough to prevent global extinction is debatable and will depend on the interplay between different biological and abiotic feedbacks. Still, models incorporating these basic principles predict that, on average, the presence of a biosphere will tend to increase planetary habitability.

With a sample size of just one Earth it is difficult (some would say impossible) to test whether the generally expected result of abundant life on a planet is to increase or decrease its habitability. However, if remote life on an exoplanet is indeed detected in the next few decades, we will begin to build up a larger sample size of inhabited worlds. With progressively more advanced space telescopes future scientists could resolve more about the properties of these worlds, and contrast them with the properties of exoplanets in the habitable zone that show no signs of being inhabited. In this way we could finally test the Gaia hypothesis. Whatever the result, we would learn something profound about the nature of inhabited planets in general, not just the Earth in particular.

Exo-Earth system science

Climate dynamics, geology, and biology are all causally intertwined in the Earth system, and are bound to be intertwined on other inhabited worlds. By generalizing our models of the Earth system and its development, researchers are beginning to formulate what I would christen 'Exo-Earth system science'—a general science of habitable and inhabited worlds. In the next decade we will begin to be able to test the predictions of those models against new observations of what our current theories say should be potentially habitable exoplanets. There are bound to be surprises—perhaps profound ones—about the prevalence of habitable worlds, and of life, in the nearby cosmos. Perhaps we will find that despite all those potentially habitable planets out there, there are no signs of life on any of them. Perhaps we will find abundant life and will be left pondering why after fifty years of searching, no signal of extraterrestrial intelligence has been detected. Either way, the results are destined to change our view of ourselves and of our world. I trust that we will look back at the Earth and our own intelligence with a renewed sense of wonder and a determination to help sustain this remarkable planet.

References

The data in Figure 21 are from the Carbon Dioxide Information
    Analysis Center—Tom Boden, Bob Andres, and Gregg Marland.
The data in Figure 22 are from Dr Pieter Tans, ESRL/NOAA, and
    Dr Ralph Keeling, Scripps Institution of Oceanography.
Figure 27 is based on Steffen et al. (2015) 'Planetary Boundaries:
    Guiding Human Development on a Changing Planet'. *Science*
    347: 736.
Figure 30 is based on results from Kopparapu et al. (2013) 'Habitable
    Zones around Main-Sequence Stars: New Estimates'. *Astrophysical
    Journal* 765: 131, and the habitable zone calculator from the NASA
    Astrobiology Institute's Virtual Planetary Laboratory.
Figure 31 is adapted from Quintana et al. (2014) 'An Earth-Sized
    Planet in the Habitable Zone of a Cool Star'. *Science* 344: 277–80.

Further reading

Chapter 1: Home

James Lovelock, *Gaia: A New Look at Life on Earth* (Oxford University Press, 1979). A popular science classic—Lovelock's poetic description of his dawning realization of Gaia.

Earth System Sciences Committee, NASA Advisory Council, *Earth System Science Overview: A Program for Global Change* (NASA, 1986). A founding document for a new scientific field.

Martin Redfern, *The Earth: A Very Short Introduction* (Oxford University Press, 2003). A succinct introduction to the workings of the inner Earth.

Chapter 2: Recycling

Lee R. Kump, James F. Kasting, and Robert G. Crane, *The Earth System*, 3rd edition (Prentice Hall, 2010). An introductory textbook to the field, nicely rooted in a systems thinking approach.

Michael C. Jacobson, Robert J. Charlson, Henning Rodhe, and Gordon H. Orians (eds), *Earth System Science: From Biogeochemical Cycles to Global Change* (Academic Press, 2000). A more advanced textbook covering key aspects of Earth system science.

William H. Schlesinger and Emily S. Bernhardt, *Biogeochemistry: An Analysis of Global Change*, 3rd edition (Academic Press, 2013). A more advanced textbook focused on biogeochemical cycling.

Chapter 3: Regulation

James Lovelock, *The Ages of Gaia: A Biography of Our Living Earth* (Oxford University Press, 1988). My favourite popular science book. Lovelock transforms Gaia into a theory using simple models to explore self-regulation, and laying out the history of our planet as a system.

Chapter 4: Revolutions

Tim Lenton and Andrew Watson, *Revolutions that Made the Earth* (Oxford University Press, 2011). Our synthesis of the pivotal events that created a world in which humans could evolve.

Charles H. Langmuir and Wally Broecker, *How to Build a Habitable Planet: The Story of Earth from the Big Bang to Humankind* (Princeton University Press, 2012). Long but rewarding—a comprehensive introduction to the Earth in a universal context.

Paul G. Falkowski, *Life's Engines: How Microbes Made Earth Habitable* (Princeton University Press, 2015). An insightful and personal introduction to the organisms that really run this planet.

Donald E. Canfield, *Oxygen: A Four Billion Year History* (Princeton University Press, 2014). The pivotal gas for complex life gets a welcome biography from one of the leaders of the field.

Chapter 5: Anthropocene

Jan Zalasiewicz, *The Earth After Us* (Oxford University Press, 2008). Both a neat introduction to geology and a long-term view of our legacy in the rock record.

Jared Diamond, *Guns, Germs and Steel: A Short History of Everybody for the Last 13,000 Years* (Vintage, 1998). A classic exposition of the differing development of societies over the Holocene.

Jared Diamond, *Collapse: How Societies Choose to Fail or Survive* (Penguin, 2011). A provocative look at why past civilizations came to grief and what we can learn from their mistakes.

Chapter 6: Projection

John Houghton, *Global Warming: The Complete Briefing*, 5th edition (Cambridge University Press, 2015). The definitive summary of the complexities of climate change.

James R. Fleming, *Historical Perspectives on Climate Change* (Oxford University Press, 1998). A myth-busting series of essays on the history of thinking about climate change.

Kendal McGuffie and Ann Henderson-Sellers, *The Climate Modelling Primer*, 4th edition (Wiley-Blackwell, 2014). Makes the fiendishly complex business of climate modelling accessible.

David Archer, *The Long Thaw: How Humans are Changing the Next 100,000 Years of Earth's Climate* (Princeton University Press, 2010). The long view on the consequences of fossil fuel burning.

Chapter 7: Sustainability

Donella H. Meadows, Dennis L. Meadows, Jørgen Randers, and William W. Behrens III, *The Limits to Growth* (Universe Books, 1972). A bible for the environmental movement but a heresy for many economists—either way, the systems modelling approach taken was groundbreaking.

David J. C. Mackay, *Sustainable Energy—Without the Hot Air* (UIT, 2009). A quantitative examination of what it will take to power our civilizations sustainably.

Julian M. Allwood and Jonathan M. Cullen, *Sustainable Materials—With Both Eyes Open* (UIT, 2012). A careful look at how we can be more efficient with key materials.

Chapter 8: Generalization

James Kasting, *How to Find a Habitable Planet* (Princeton University Press, 2012). The maestro of modelling habitable zones offers the perfect manual for the would-be planet finder.

David C. Catling, *Astrobiology: A Very Short Introduction* (Oxford University Press, 2013). A sparkling introduction to the prospects for life elsewhere.

Index

Note: **bold** font indicates appearance in a figure

Index

SOCIAL MEDIA
Very Short Introduction

Join our community
www.oup.com/vsi

- Join us online at the official Very Short Introductions **Facebook** page.
- Access the thoughts and musings of our authors with our online **blog**.
- Sign up for our monthly **e-newsletter** to receive information on all new titles publishing that month.
- Browse the full range of Very Short Introductions online.
- Read **extracts** from the Introductions for free.
- Visit our library of **Reading Guides**. These guides, written by our expert authors will help you to question again, why you think what you think.
- If you are a teacher or lecturer you can order inspection copies quickly and simply via our website.